SpringerBriefs in Plant Science

SpringerBriefs present concise summaries of cutting-edge research and practical applications across a wide spectrum of fields. Featuring compact volumes of 50 to 125 pages, the series covers a range of content from professional to academic. Typical topics might include:

- A timely report of state-of-the art analytical techniques
- A bridge between new research results, as published in journal articles, and a contextual literature review
- A snapshot of a hot or emerging topic
- An in-depth case study or clinical example
- A presentation of core concepts that students must understand in order to make independent contributions

SpringerBriefs in Plant Sciences showcase emerging theory, original research, review material and practical application in plant genetics and genomics, agronomy, forestry, plant breeding and biotechnology, botany, and related fields, from a global author community. Briefs are characterized by fast, global electronic dissemination, standard publishing contracts, standardized manuscript preparation and formatting guidelines, and expedited production schedules.

More information about this series at http://www.springer.com/series/10080

M. Ali • Khalid Rehman Hakeem

Scientific Explorations of *Adhatoda vasica*

An Asian Health Remedy

 Springer

M. Ali
College of Pharmacy, Department
of Pharmacognosy
Jazan University
Jazan, Saudi Arabia

Khalid Rehman Hakeem
Department of Biological Sciences, Faculty
of Science
King Abdulaziz University
Jeddah, Saudi Arabia

ISSN 2192-1229   ISSN 2192-1210   (electronic)
SpringerBriefs in Plant Science
ISBN 978-3-030-56714-9   ISBN 978-3-030-56715-6   (eBook)
https://doi.org/10.1007/978-3-030-56715-6

This Springer imprint is published by the registered company Springer Nature Switzerland AG
The registered company address is: Gewerbestrasse 11, 6330 Cham, Switzerland

(1908–1999)

*A great philanthropist, thinker, visionary, and founder-chancellor of **Jamia Hamdard**, New Delhi, India*

Preface

The therapeutic potential of phytoremedies to cure diseases is due to the presence phytoconstituents in different parts of a plant. Since the ancient times, traditional plant medicine has been a "hit and trial" approach that was memorized and transmitted verbally from one generation to the next. Infact the plant medicines or botanicals have drawn people's attention in all eras and are used in different systems of medicine (Ayurveda, Unani, Siddha, and Homeopathy). Since medicinal plants have a great potential to combat diseases, it becomes more important to understand those plants through scientific approaches considering their origin and modern updates. This book deals with the scientific details of *Adhatoda vasica* L. Nees, which is a promising and potential medicinal plant and is indigenous to India and abundantly found in Asian regions. The plant consists of mainly alkaloids, but secondary metabolites in different parts of the plant exhibit seasonal quantitative variations.

The chapters of the book emphasize on various aspects of the plant such as its profile, geographical distribution, and taxonomical details. Phytochemical and biotechnological updates in this book reveal the plant as an important remedy to cure various diseases of different physiological systems. Though the plant constituents favor the treatment of pulmonary diseases, fractional extracts possess a wide range of therapeutic efficacy. Ethnomedicine and its pharmacological aspects, as well as their recent updates, are the most attractive part of the book. The book also elaborates perceptions about the existing formulations for various ailments and suggests the possibilities of future work that could be designed to explore any hidden dynamicity. Patent details regarding the isolation of various plant components and their dosage forms are summarized in a separate chapter. This book will be useful for both students and research scholars in this field.

Jazan, Saudi Arabia   M. Ali
Jeddah, Saudi Arabia   Khalid Rehman Hakeem

Acknowledgement

We acknowledge the support of **Dr. Shah Alam Khan** (Professor, College of Pharmacy, National University of Science and Technology, Muscat, Sultanate of Oman) for contributing patent details pertaining to various aspects of the medicinal plant *Adhatoda vasica* in the book. Prof. Khan is a magnanimous personality and an expert in pharmacological subjects. His guidance and contribution has surely increased the value of this book.

Contents

About the Authors

Dr. M. Ali is currently working under Ministry of Higher Education as a lecturer of Pharmacognosy in the department of Pharmacognosy, College of Pharmacy, Jazan University, Kingdom of Saudi Arabia (KSA). He received his Ph.D. from Jamia Hamdard University, New Delhi, India in 2012 for "Phytochemical and Pharmacological studies of an Anticancer Medicinal Plant and its Authentication using Molecular Biology Techniques". He is also recipient of Government of India, Junior research fellowship (JRF, GATE) fellowship for pursuing master's in pharmacy (M. Pharm) for "Enhanced Production of Vasicine and Vasicinone from callus culture of *Adhatoda vasica*". He has published/presented about 25 research articles in the national and international journal of repute. He has reviewed scientific papers in field of Pharmaceutical science.

Prof. (Dr.) Khalid Rehman Hakeem, PhD, is Professor at King Abdulaziz University, Jeddah, Saudi Arabia. After completing his doctorate (Botany, specialization in Plant Eco-physiology and Molecular Biology) at Jamia Hamdard, New Delhi, India, in 2011, he worked as a Lecturer at the University of Kashmir, Srinagar, for a short period. Later, he joined Universiti Putra Malaysia, Selangor, Malaysia, and worked there as Postdoctorate Fellow in 2012 and Fellow Researcher (Associate Prof.) from 2013 to 2016. Dr. Hakeem has more than 10 years of teaching and research experience in plant eco-physiology, biotechnology and molecular biology, medicinal plant research, plant-microbe-soil interactions, as well as in environmental studies. He is the recipient of several fellowships at both national and international levels; he has also served as a Visiting Scientist at Jinan University, Guangzhou, China. Currently, he is involved in a number of international research projects with different government organizations.

So far, Dr. Hakeem has authored and edited more than 50 books with international publishers, including Springer Nature, Academic Press (Elsevier), and CRC Press. He also has to his credit more than 110 research publications in peer-reviewed international journals and 60 book chapters in edited volumes with international publishers.

At present, Dr. Hakeem serves as an editorial board member and reviewer of several high-impact international scientific journals from Elsevier, Springer Nature, Taylor and Francis, Cambridge, and John Wiley Publishers. He is included in the Advisory Board of Cambridge Scholars Publishing, UK. He is also a Fellow of Plantae Group of the American Society of Plant Biologists; Member of the World Academy of Sciences; Member of the International Society for Development and Sustainability, Japan; and Member of the Asian Federation of Biotechnology, Korea. Dr. Hakeem has been listed in Marquis Who's Who in the World since 2014. Currently, Dr. Hakeem is engaged in studying the plant processes at eco-physiological as well as molecular levels.

Chapter 1
Introduction

Traditional systems of medicines have always been an attention-seeking aspect of all the mode of practices to cure the various ailments since thousands of years. The coexistence of plant and the rest of the creations (including human beings) are like the complimentary strands of DNA as necessity. The existence of one decides the surviving of the rest. It's not possible to determine the exact historical dates to refer to the use of plants, but some evidences support the claim of the cultivated drugs approximately sixty thousands (60,000) years back incredibly (Solecki and Shanidar 1975), while at the other hand the scripts pertaining to the medicinal plants date back to around 5000 years in the regions as Egypt, India and China while approximately 2500 years in Greece and Central Asia (Ang-Lee et al. 2001). Asian regions are blessed with tremendous flora where *Adhatoda vasica* is one of them a promising and well documented in ancient literature. The ancient time was bound to adopt the 'trial and error' approach to use the natural products and various crude extracts to treat the patients, and gradually human beings were familiar to apply the positive outcomes to meet the curable requirements (Jamshidi-Kia et al. 2018). Ethno-pharmacological evidences of the said plant are based on regional and traditional practices.

The contribution for the traditional approaches to treat the ailments would be incomplete without mentioning the Ancient Greek people who were known to have the sound knowledge of local medicinal herbs and various other plants. Hippocrates as being the Greek origin was known as the founder of Greek medicine. His pupil Theophrastus learned a lot, and both used medicinal plants as excellent health remedies. The sequential efforts were in progress and various literature works were substantially documented as encyclopaedia to describe several hundreds of medicinal plants with overwhelming therapeutic efficacies adopting the scientific modes of studies. The encyclopaedia *De Materia Medica* was written (75–45 BC) by Pedanius Dioscorides (Lindberg and Bertelsen 1995; Rios and Recio 2005; Zargari 1992).

M. Ali, K. R. Hakeem, *Scientific Explorations of Adhatoda vasica*,
SpringerBriefs in Plant Science, https://doi.org/10.1007/978-3-030-56715-6_1

The lack of technical prowess, facilities, and rationalized protocols of all the traditional practices compelled the practitioners to memorize the way to treat till the complete cure because the use of medicinal plants was the only option as the way of treatment (Halberstein 2005). The present book embarks upon the Asian health remedy (*A. vasica*) as a gift of nature having innumerable phytochemical constituents, i.e. alkaloids, flavonoids, volatile oils and traces of other metabolites. The presence of metabolites is the sole foundation for designing the drugs and formulations. Sometimes the natural complexities and physical intricacies make it quite cumbersome to streamline the work practice and proper documentations. Now in the twenty-first century, the human beings are competent enough to their technical prowess to develop and to process the medicinal plants to its proper destiny it deserves. The therapeutic potential of natural products has been considered a fruitful constructive and promising approach to manage the health-care system and its advancements.

The twenty-first centuries provided a plenty of scientific explorations and advancements to discover the chemical behaviour of bioactive compounds from natural herbs and other medicinal plants. In the beginning, it was so difficult to finalize the establishment of molecular structures of the extracted or isolated compounds. But the gradual timings and scientific attentions were enough to develop the highly précised chromatographic and spectroscopic instruments such as high-performance liquid chromatography (HPLC), high-performance thin layer chromatography (HPTLC), UPLC, UV, IR, NMR, mass spectroscopies, etc. The advancements were not confined on the single applications, but the coupled scientific instrumentation approaches were adopted to enhance the precision and to save the time. Examples include LC-MS, GC-MS, IC-MS, etc. which are meant for quantitative and qualitative applications being helpful for standardization of crude materials which is the most significant issue for phytolabs and industries (Yadav et al. 2014). The phytochemical analysis of *A. vasica* has been completed on all stages of screenings which was more important to design the better derivatives and formulations to treat the ailments with single or combined compositions. All parts of the *A. vasica* plant contain the unique features in terms of bioactive principles. This diversity among metabolites is responsible to vindicate the usefulness of plant to control and cure the various physiological and pathological disorders. Therapeutic potential is also attributed to the varying polarities of the constituents and their nature.

The present book reveals the inclusivity about the *A. vasica* plant for its Asian distribution on plain and high altitude locations, cultivation pattern and climatic conditions, profile to identify it on the basis of pharmacognostical (macroscopical and microscopical) standardizations studies, bioactive profiles, ethnopharmacological importance and evidences for Indian traditional systems (AYUSH). The utilization of *A. vasica* plant is highly exploited to meet the requirements of phytochemical constituents to develop the remedies and research activities. The use of biotechnological approaches to conserve its germplasm for future propagation is the attractive feature of manuscript. The biotransformation studies and conditions for micro-propagation have been summarized as guidance to execute the research activities to fulfil the demands when the plants are not available in enough amounts naturally. *A. vasica* with its commercial importance is also mentioned to show its necessity to promote the employment and economic contributions.

References

Ang-Lee MK, Moss J, Yuan CS (2001) Herbal medicines and perioperative care. JAMA 286(2):208–216

Halberstein RA (2005) Medicinal plants: historical and cross-cultural usage patterns. Ann Epidemiol 15(9):686–699

Jamshidi-Kia F, Lorigooini Z, Amini-Khoei H (2018) Medicinal plants: past history and future perspective. J Herb Pharmacother 7(1):1–7

Lindberg MH, Bertelsen G (1995) Spices as antioxidants. Trends Food Sci Technol 6(8):271–277

Rios JL, Recio MC (2005) Medicinal plants and antimicrobial activity. J Ethnopharmacol 100(1–2):80–84

Solecki R, Shanidar IV (1975) A Neanderthal flower burial in Northern Iraq. Science 190(4217):880–881

Yadav M, Chatterji S, Gupta SK, Watal G (2014) Preliminary phytochemical screening of six medicinal plants used in traditional medicine. Int J Pharm Pharm Sci 6(5):539–542

Zargari A (1992) Medicinal plants. Tehran University Press, p 889

Chapter 2
Distribution and Availability of the Present Remedy

Adhatoda vasica is indigenous to India and is available at various locations ranges from plain areas to the mountains. The plant availability in Himalayan region reveals the altitudinal range from 200 to 1300 m (Negi et al. 2018; Gantait and Panigrahi 2018). The geographical conditions having different intensities of biotic and abiotic factors are responsible to bring the genetic and physiological changes at molecular level in plant development almost for all the species (Atkinson and Urwin 2012). Morphoanatomical and the variation in chemical profiling are the attributes of biotic and abiotic factors (Fabio Cassola et al. 2019).

The tropical regions of Southeast Asia are considered to have *A. vasica* plant with its some other species. The geographical conditions and the climatic factors are found much compatible for the better growth and cultivation of vasaka. To be more specific as the location is concern, India (indigenous) and the neighbouring countries as Pakistan, Bangladesh, Afghanistan, Bhutan, Malaya and Sri Lanka having all the climatic seasons are more important to have the plant. The propagation is not confined to only the mentioned countries but well extended via China to Hong Kong, Taiwan, Ethiopia, Cyprus, etc. and particular sites in Sweden and Germany as well. Kokate et al. (2010) mentioned its cultivation to be non-commercial and stated about its availability through garden and wild sources. Seed germination and stem cuttings with 3–4 nodes are the means of ideal propagation or planting and can be obtained throughout all the seasons as its indigenous locations are most preferably in loamy and alluvial soils. Optimum climatic conditions (a soothing weather) are best suited for early nursery preparations then to be transferred in the field. Pure and intercrop cultivations are the choices to grow the plant to suit the purpose. Root cutting requires to be planted on ridges having 55–60 cm gap and the 25–30 cm within the cuttings itself. Water logging must be monitored to avoid any deleterious effect of planted cuttings. The collection of the plant part is a crucial step to suit the purpose and hence must be chosen around January due the certified facts of

M. Ali, K. R. Hakeem, *Scientific Explorations of Adhatoda vasica*,
SpringerBriefs in Plant Science, https://doi.org/10.1007/978-3-030-56715-6_2

containing the maximum metabolites (*http://www.celkau.in/Crops/Medicinal%20 Plants/Adathoda.aspx*).

References

Atkinson NJ, Urwin PE (2012) The interaction of plant biotic and abiotic stresses: from genes to the field. J Exp Bot 63(10):3523–3543

Cassola F, da Silva MHR, Borghi AA, Lusa MG, Sawaya ACHF, Garcia VL, Mayer JLS (2019) Morphoanatomical characteristics, chemical profiles, and antioxidant activity of three species of Justicia L.(Acanthaceae) under different growth conditions. Ind Crop Prod 131:257–265

Gantait S, Panigrahi J (2018) In vitro biotechnological advancements in Malabar nut (*Adhatoda vasica* Nees): Achievements, status and prospects. J Genet Eng Biotechnol 16(2):545–552

Kokate CK, Purohit AP, Gokhle SB (2010) Pharmacognosy, 46, Nirali Prakashan Pune, 3(2) pp 3.84–3.86

Negi VS, Kewlani P, Pathak R, Bhatt D, Bhatt ID, Rawal RS, Sundriyal RC, Nandi SK (2018) Criteria and indicators for promoting cultivation and conservation of medicinal and aromatic plants in Western Himalaya, India. Ecol Indic 93:434–446

Chapter 3
Plant Profile and Documented Evidence in Different Systems of Traditional Medicine

3.1 Description, Identity and Taxonomy

Adhatoda vasica L. Nees (Malabar nut) belongs to the family Acanthaceae, known by different names in scientific binomial system and vernacular section. It is an evergreen shrub having not any pleasant smell or flavour but a prevailing sense of bitterness. Maximum height reaches up to 2.5 m. Ancient document confirms its authenticity as well-known expectorant remedy in both the systems of Ayurvedic and Unani (Chopra et al. 1956; Kapoor 2001). In northeastern part of India, the Naga tribes have a long back history to use the decoction of the leaf of *A. vasica*, locally known as 'sorukni' a traditional medicinal to get rid of intestinal infections due to worms (Yadav and Tangpu 2008). The vernacular names are tabulated in Table 3.1 in different languages. These are some selective names, while others have been figure out in digits as Hindi (24), Kannada (25), Marathi (11), Sanskrit (57), Tamil (89), Telugu (12), Malayalam (6), Tibetan (4) and Urdu (7) (http://envis.frlht. org/plantdetails/46e27646a0964750a598ed5b2041f86e/44ced0a0aa90fdb5399d93 bc8481107e).

Genus *Adhatoda* is very huge to contain approximately 157 species. These are all distributed in the different continents in the world but '*vasica*' is mainly found in Asian regions. The accepted 30 names of the different species have been summarized in the Table 3.2, while *vasica* species has been taxonomically classified and mentioned in Table 3.3.

M. Ali, K. R. Hakeem, *Scientific Explorations of Adhatoda vasica*, SpringerBriefs in Plant Science, https://doi.org/10.1007/978-3-030-56715-6_3

Table 3.1 Vernacular names of *Adhatoda vasica* in different Indian languages

English	Adhatoda
Kannada	Baesaala, aadsoge, adusage, aadu sokada gida, addasara, byalada, edmuttandittapu, yedumuttanditappu
Marathi	Adulsa, arusa, baksa, adulso
Sanskrit	Vasaka, adarushah, bhisahkmata, bhishagmata, vaidyamatruvrikshaha, vaidyasinhi, vrsah, vrsaka
Tamil	Adathodai, adatoda, attakani, attakanicceti, atutotai, calamaram catikacceti, catikam, akacattamarai, antarati tamarai
Telugu	Adda saramu, adda-sarap, kaatumurungai, vachai, adasaram, addasaram
Malayalam	Adel-odagam, atalotakam, ataloetakam, attalotakam
Tibetan	Ba sa ka, ba-sa, ba-sa-ka, bri-sa
Hindi	Arusha, asgand, bansa, adusa, vasak, adosa, adulasa, basute, basuti
Urdu	Adoosa, arusa (bansa), bansa ke pattay, burg arusa, burg bansa, burg-i-arusa

3.2 Pharmacognostic Profile

A. vasica bears short petiolate leaves (size 12–20 cm long, 2.6–6.0 cm broad), glabrous surface with perfect elliptical lanceolate, entire and slight crenate margin, opposite leaves (Fig. 3.1) with oval diacytic stomata, pinnate venation with prominent midrib and acuminate apex with tapering base; upon drying it tastes bitter and brownish green in colour (Singh et al. 2011). Flowers are of white, pink or purple colour turning brownish dull upon drying. Fruits and seeds are globular in nature (Patel and Venkatakrishna-Bhatt 1984; Kumar et al. 2013).

Histological studies include its transverse section and powder studies which show upper and lower epidermis with slight thin-walled polygonal cells of wavy margin. Stomata can easily be observed in TS and more clearly in leaf powder (caryophyllaceous type); the leaf is dorsiventral having the palisade cells at one surface of upper epidermis but in double layer. Uniseriate trichomes are found at both the surfaces having the description of multicellular, covering and few glandular as well. Five to six layers of cells are comprised by spongy parenchyma. Collenchymatous cells are present at both the epidermis surfaces; medullary rays are also present. Mid rib region also contains oil globules and crystals of calcium carbonate and calcium oxalate (acicular and prismatic) types as given in Fig. 3.2 (Kokate et al. 2010).

3.3 Bioactive Profile of *A. vasica*

The variety and wide range of secondary metabolites i.e. Quinazoline alkaloids – vasicine and vasicinone, non-crystalline steroid vasakin, glycosides, phenolic compounds, sterols, fatty acids and essential oils (The Wealth of India 2003) in the different parts of the plant, the therapeutic potential is shown to cure or alleviate the

Table 3.2 *Adhatoda* species with their global biodiversity and geographical locations

S. No.	Species for genus *Adhatoda*	Global biodiversity	
		Accepted name	Location
1.	*Adhatoda acuminata* Nees	*Justicia flava* Vahl	Ethiopia
2.	*Adhatoda arenaria* Nees	*Justicia trinervia* (Nees)	India
3.	*Adhatoda capensis* Nees	*Adhatoda capensis* var. *glabrescens* Nees	South Africa
4.	*Adhatoda congrua* Nees	*Adhatoda congrua* Lindau Nees	Brazil
5.	*Adhatoda cuneifolia* Nees	*Adhatoda cuneifolia* Hook. ex Nees	Brazil
6.	*Adhatoda densiflora* (Hochst.) J.C. Manning	*Adhatoda densiflora* (Hochst.) J.C. Manning	South Africa
7.	*Adhatoda engleriana* (Lindau) C.B.Clarke	*Adhatoda engleriana* Lindau	Kenya, Tanzania, Uganda,
8.	*Adhatoda eustachyana* Nees	*Adhatoda eustachyana* Jacq	Graz
9.	*Adhatoda gilliesii* Nees	*Adhatoda gilliesii* Griseb. & Hook.	Argentina
10.	*Adhatoda holosericea* Nees	*Adhatoda holosericea* Nees	Brazil
11.	*Adhatoda hyssopifolia* Nees	*Adhatoda hyssopifolia* Nees	Spain
12.	*Adhatoda le-testui* (Benoist) Heine	*Duvernoia le-testui* (Benoist)	Gabon, Congo
132.	*Adhatoda leptantha* (Nees) Nees	*Justicia tubulosa*	Mozambique
14.	*Adhatoda leptostachya* Nees	*Justicia caliculata* Deflers	Yemen
15.	*Adhatoda lindeniana* Nees	*Adhatoda lindeniana* Nees	Venezuela, Colombia
16.	*Adhatoda natalensis* Nees	*Justicia natalensis* T. Anderson	South Africa
17.	*Adhatoda nuda* Nees	*Justicia kuntzei*	Yemen
18.	*Adhatoda odora* (Forssk.) Nees	*Justicia odora* Vahl	Saudi Arabia
19.	*Adhatoda orbicularis* (Lindau) C.B.Clarke	*Duvernoia orbicularis* Lindau	Cameroon
20.	*Adhatoda orbigniana* Nees	*Thyrsacanthus boliviensis* Nees	Bolivia
21.	*Adhatoda orchioides* Nees	*Justicia orchioides*	South Africa
22.	*Adhatoda palustris* (Hochst.) Nees	*Justicia palustris* T. Anderson	Eritrea
23.	*Adhatoda protracta* (Nees) Nees	*Justicia protracta* T. Anderson	South Africa
24.	*Adhatoda reflexiflora* (Vahl) Nees	*Justicia periplocifolia* Jacq.	Puerto Rico, India
25.	*Adhatoda tetramera* Bello	*Justicia sessilis* (Jacq.)	Unknown
26.	*Adhatoda thymifolia* Nees	*Justicia thymifolia* C.B. Clarke	South Africa
27.	*Adhatoda tristis* Nees	*Justicia tristis* T.Anderson	Africa, Cameroon
28.	*Adhatoda tubulosa* Nees	*Justicia tubulosa*	Unknown
29.	*Adhatoda tweediana* Nees	*Poikilacanthus tweedianus* Lindau	Panama

(continued)

Table 3.2 (continued)

S. No.	Species for genus *Adhatoda*	Global biodiversity	
		Accepted name	Location
30.	*Adhatoda ventricosa* Nees	*Adhatoda ventricosa* Wall. Ex Sims.	China

Courtesy – http://www.plantlist.org/tpl1.1/search?q=adhatoda

Table 3.3 Taxonomical classification of *Adhatoda vasica* L. Nees

Kingdom	Plantae
Subkingdom	Tracheobionta – vascular plant
Super division	Spermatophyta – seed plants
Division	Magnoliophyta
Class	Magnoliopsida
Subclass	Asteridae
Order	Scrophulariales – Lamiales
Family	Acanthaceae – *Acanthus* family
Genus	Adhatoda
Species	*vasica*

diseases of respiratory system (i.e. asthma, cold, *Mycobacterium tuberculosis* by producing ambroxol and bromhexine (Narimaian et al. 2005) and cough (Sharma et al. 1992), Bronchodilatation (vasicine), bronchoconstriction (vasicinone), expectorant and antispasmodic (Karthikeyan et al. 2009). Cardiac effect (a combination of vasicine and vasicinone shows the significant reduction in cardiac depressant), anticholinesterase (Lahiri and Pradhan 1964), antidiabetic ((Gao et al. 2008), anti-ulcer activity (tested in rats) when compared positively with the actions of pylorus and aspirin drugs (Shrivastava et al. 2006), cardiovascular protection, abortifacient, antimutagenic, antitubercular, antiallergic, anti-inflammatory, oxytocic and anti-asthmatic, antitussive and hepatoprotective (Singh et al. 2011). The leaves syrup prepared from *A. vasica* improved the symptoms of dyspepsia (Chaturvedi et al. 1983). Plant extracts showed antimutagenic activity upon cadmium-intoxicated mice (Jahangir et al. 2006) and radioprotective effects on testis (Kumar et al. 2007).

European practitioners have also used *A. vasica* for various reasons by taking its fluid extract and tincture as an antispasmodic, febrifuge and expectorant, typhus fever and diphtheria. Significant appreciation has been accounted in Germany where the leaves are employed as an expectorant and spasmolytic and for the remedy of common cold, bronchitis, laryngitis, influenza, cramp, pertussis, dry cough, hay fever, sinusitis and asthma (Claeson et al. 2000). Some additional properties of the extracts of different parts of the *A. vasica* are alphabetically enumerated in the Fig. 3.3. A very popular Bengali saying covers the community's respect for *A. vasica* plant, translating as, 'a man wouldn't die of disease in the region where the plants *Vitex negundo* (bana), *Acorus calamus* (calamus) and *Adhatoda vasica* (vasaka) are present' provided that the user knows how to use them (Uniyal et al. 2006).

Fig. 3.1 *Adhatoda vasica* LNees (unpublished original image)

3.4   Traditional Systems of Medicine

3.4.1   Ayurveda

The plant is known with many synonyms having importance in Ayurvedic system of medicine for the treatment of ailments pertaining to the respiratory tract in both adults and children (Hossain and Hoq 2016). It is quite significant to have more than 20 formulations utilizing it in the form of juices. Being always in demand and possessing compliance with desirable results made it to be included in the manual prescribed by World Health Organization. The use of the plant as the traditional medicine in primary health care drew the attention of field researchers (Kapgate and Patil 2017). There was an ancient Indian saying that, 'No man suffering from phthisis needed despair as long as the vasaka plant (A. vasica) exists' (Dymock et al. 1893). It has been used in Ayurvedic system of medicine. In Ayurvedic preparations, vasaka leaf juice (*Vasa swarasa*) is as the important ingredient which is incorporated in more than 20 polyherbal formulations which include Vasarishta, Mahatiktaka Ghrita, Triphala ghrita, Vasakasava, Vasavaleha, Mahatriphalaghrita, Panchatikta ghrita and Panchatiktaghritaguggulu (Anonymous 2000). Charaka Samhita has categorized the drug under separate section of mucolytic and expectorant drugs. Properties and actions in Ayurveda have its own terminology as Rasa- Tikta, Kasaya, Guna – Laghu, Virya – Sita, Vipaka- Katu, Karma – Hrdya, Kaphapittahara, Raktasangrahika, Kasaghna (Anonymous 2008).

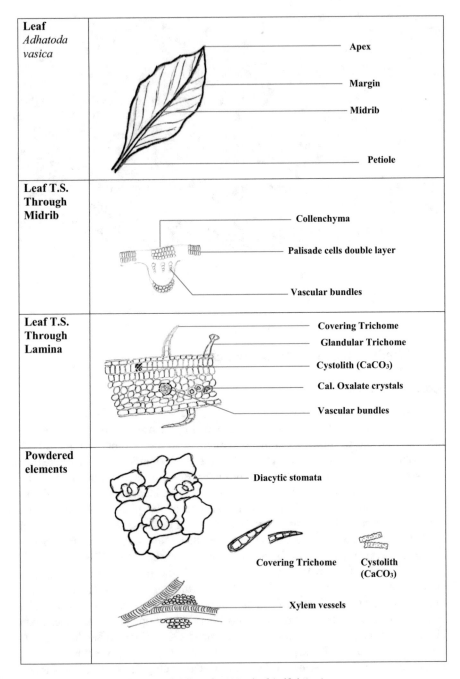

Fig. 3.2 Microscopical features of *Adhatoda vasica* leaf (self-drawn)

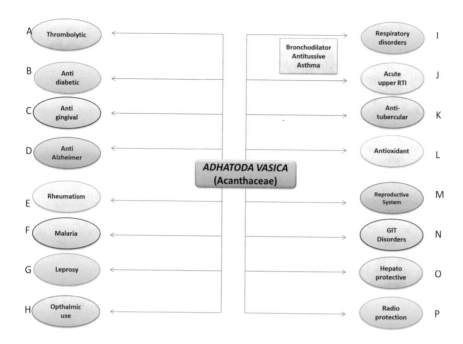

Fig. 3.3 The medicinal potential of *Adhatoda vasica* in different extractive modes
Figure references (*A*- Fahad et al. 2014, *B*- Gao et al. 2008, *C*-Jayashankar et al. 2011, *D*-Shereen et al. 2013, *E–H*- Atta-Ur-Rahman et al. 1986, *I*- Amin and Mehta 1959; Johri and Zutshi 2000, Paliwa et al. 2000, *J*- Narimaian et al. 2005, *K*- Gupta et al. 2010, *L*- Jahangir and Sultana 2007, *M*- Burgos et al. 1997, *N*- Bharti et al. 1995, *O*- Bhattacharyya et al. 2005, *P*- Kumar et al. 2005)

3.4.2   Siddha

Siddha is the ancient Indian traditional system originated in South India and is based on a concept of medicinal practices and spiritual mode of disciplines. *Adhatoda vasica* is known as *Adathodai ilai*, and the formulation mode identify it as *Adathodai ilai* chooranam; it is a Siddha preparation. Siddha preparation is meant for the treatment of hypertension (Selvakumar et al. 2018).

Siddha formulations are having *A. vasica* as the important ingredient in several home remedies. Along with other ingredients, i.e. Thippili (*Piper longum*), pepper, chukka (dry ginger) and Thoothuvalai (*trilobatum*), *A. vasica* is commonly used in treating the respiratory ailments and asthma in particular. Tested Siddha remedies also include Adathodai kudineer (decoction by boiling) and Manappagu for asthma. Flowers can also be used for ophthalmic purposes to cure eye irritation through the Siddha protocols of heating for a short while (Kumar et al. 2010).

3.4.3   Unani

Atal 1980 mentioned about the system of Unani medicine (Anonymous 1948) in practice for more than 2000 years which includes *A. vasica* plant as the important drug in the indigenous system of medicine in Indian history. The Unani formulations having *A. vasica* as the important ingredient are Safoof Arusa, Gulqand, Sharbat-e- Aijaz, Habb-e- Zeeq-un- Nafas, etc.

3.4.4   Homeopathy

Homeopathy is also an alternative system of treatment where very low doses of natural components are employed on the principle of 'like cures like'. The presence of phytoconstituents in *A. vasica* at nanosize scale with variable potencies of 6c and 30c is prepared to combat the respiratory disorders. The significant rationale and validity of this alternative therapy is always surrounded with controversy but continues to flourish because of its effectiveness to fight the diseases (Mandal et al. 2019).

References

Amin AH, Mehta DR (1959) A bronchodilator alkaloid (vasicinone) from *Adhatoda vasica* nees. Nature 184(Suppl 17):1317

Anonymous (1948) Wealth of India, vol 1. C.S.I.R. Department of Scientific Research, Govornment of India, p 253

Anonymous (2000) Database on medicinal plants used in ayurveda, vol I. Central Council of Research in Ayurveda and Siddha, Department of Indian System of Medicine and Homeopathy, Ministry of Health and Family Welfare (Govornment of India), New Delhi, pp 496–509

Anonymous (2008) The ayurvedic pharmacopoeia of India. Part-I Volume-VI, 1st edn, pp 161–162

Atal CK (1980) Chemistry and pharmacology of vasicine: a new oxytocin and abortifacient. Indian Drugs 15:15–18

Atta-Ur-Rahman, Said HM, Ahmad VU (1986) Pakistan encyclopaedia planta medica, vol 1. Hamdard Foundation Press, Karachi, pp 181–187

Bharti M, Krishna M, Tewari SK (1995) The effect of vasa (*Adhatoda vasica* nees) On amlapitta. Anc Sci Life 14:143–149

Bhattacharyya D, Pandit S, Jana U, Sen S, Sur TK (2005) Hepatoprotective activity of *Adhatoda vasica* aqueous leaf extract on D-galactosamine-induced liver damage in rats. Fitoterapia 76:223–225

Burgos R, Forcelledo M, Wagner H, Müller A, Hancke J, Wikman G et al (1997) Non-abortive effect of *Adhatoda vasica* spissum leaf extract by oral administration in rats. Phytomedicine 4:145–149

Chaturvedi GN, Rai NP, Dhani R, Tiwari SK (1983) Clinical trial of *Adhatoda vasica* syrup (vasa) in the patients of non-ulcer dyspepsia (Amlapitta). Ancient Sci Life 3(1):19

Chopra RN, Nayar SL, Chopra IC (1956) Glossary of Indian medicinal plants. Council of Scientific and Industrial Research, New Delhi

Claeson UP, Malmfors T, Wikman G, Bruhn JG (2000) *Adhatoda vasica*: a critical review of ethnopharmacological and toxicological data. J Ethnopharmacol 72(1–2):1–20

Dymock W, Warden C, Hooper D (1893) Pharmacographia India. A history of the principal drug of vegetable origin met with in British India. Kegan, Paul, Trench, Trubner and Co, London, pp 49–51

Fahad H, Ariful I, Latifa B, Mizanur RM, Mohammad SH (2014) In vitro thrombolytic potential of root extracts of four medicinal plants available in Bangladesh. Anc Sci Life 33:162–164

Gao H, Huang YN, Gao B, Li P, Inagaki C, Kawabata J (2008) Inhibitory effect on α-glucosidase by *Adhatoda vasica* Nees. J Food Chem 108(3):965–972

Gupta R, Thakur B, Singh P, Singh HB, Sharma VD, Katoch VM et al (2010) Anti-tuberculosis activity of selected medicinal plants against multi-drug resistant mycobacterium tuberculosis isolates. Indian J Med Res 131:809–813

Hossain MT, Hoq MO (2016) Therapeutic use of *Adhatoda vasica*. Asian J Med Biol Res 2(2):156–163

Jahangir T, Sultana S (2007) Tumor promotion and oxidative stress in ferric nitrilotriacetate-mediated renal carcinogenesis: protection by *Adhatoda vasica*. Toxicol Mech Methods 17:421–430

Jahangir T, Khan TH, Prasad L, Sultana S (2006) Reversal of cadmium chloride-induced oxidative stress and genotoxicity by *Adhatoda vasica* extract in Swiss albino mice. Biol Trace Element Res 111(1–3):217–228

Jayashankar S, Panagoda GJ, Amaratunga EA, Perera K, Rajapakse PS (2011) A randomised double-blind placebo-controlled study on the effects of a herbal toothpaste on gingival bleeding, oral hygiene and microbial variables. Ceylon Med J 56:5–9

Johri RK, Zutshi U (2000) Mechanism of action of 6, 7, 8, 9, 10, 12-hexahydro-azepino-[2, 1-b] quinazolin-12one-(RLX) – a novel bronchodilator. Indian J Physiol Pharmacol 44:75–81

Kapgate SM, Patil AB (2017) *Adhatoda vasica*: a critical review. Int J Green Pharm 11(4):S54–S62

Kapoor LD (2001) Hand book of ayurvedic medicinal plants. CRC Press, Boca Raton, pp 416–417

Karthikeyan A, Shanthi V, Nagasathaya A (2009) Preliminary phytochemical and antibacterial screening of crude extract of the leaf of *Adhatoda vasica* (L). Int J Green Pharm 3:78–80

Kokate CK, Purohit AP, Gokhle SB (2010) Pharmacognosy, 46, Nirali Prakashan Pune, 3(2) pp 3.84–3.86

Kumar A, Ram J, Samarth RM, Kumar M (2005) Modulatory influence of *Adhatoda vasica* nees leaf extract against gamma irradiation in Swiss albino mice. Phytomedicine 12:285–293

Kumar M, Samarth R, Kumar M, Selvan SR, Saharan B, Kumar A (2007) Protective effect of *Adhatoda vasica* Nees against radiation-induced damage at cellular, biochemical and chromosomal levels in Swiss albino mice. Evid Based Compl Alt 4(3):343–350

Kumar KPS, Bhowmik D, Chiranjib TP, Kharel R (2010) Indian traditional herbs *Adhatoda vasica* and its Medicinal application. J Chem Pharm Res 2(1):240–245

Kumar M, Dandapat S, Kumar A, Sinha MP (2013) Determination of nutritive value and mineral elements of five-leaf chaste tree (Vitex negundo L.) and Malabar nut (*Adhatoda vasica* Nees). Acad J Plant Sci 6(3):103–108

Lahiri PK, Pradhan SN (1964) Pharmacological investigation of vasicinol-alkaloid from *Adhatoda vasica* nees. Indian J Exp Biol 2(4):219

Mandal S, Mandal TK, Rath J (2019) Investigation of therapeutically active constituents of homeopathy medicine from Justicia adhatoda L. and its clinical verification. J Pharmacogn Phytochem 8(3):3790–3796

Narimaian M, Badalyan M, Panosyan V, Gabrielyan E, Panossian A, Wikman G (2005) Randomized trial of a fixed combination (KanJang) of herbal extracts containing *Adhatoda vasica*, Echinacea purpurea and Eleutherococcus senticosus in patients with upper respiratory tract infections. Phytomedicine 12:539–547

Paliwa JK, Dwivedi AK, Singh S, Gutpa RC (2000) Pharmacokinetics and in-situ absorption studies of a new anti-allergic compound 73/602 in rats. Int J Pharm 197:213–220

Patel VK, Venkatakrishna-Bhatt H (1984) In vitro study of antimicrobial activity of *Adhatoda vasica* linn. (leaf extract) on gingival inflammation – a preliminary report. Indian J Med Sci 38:70–72

Selvakumar S, Visweshwaran S, Sivakumar S, Mariappan A, Ushakanthan S (2018) Antihypertensive and diuretic action of adathodai ilai chooranam – a siddha mono-herbal formulation. Eur J Biomed 5(9):319–323

Sharma MP, Ahmad J, Hussain A, Khan S (1992) Folklore medicinal plants of Mewat (Gurgaon district), Haryana, India. Int J Pharmacogn 30(2):129–134

Shereen KA, Ahmed RH, Soltan MM, Hegazy UM, Elgorashi EE, El-Garf IA et al (2013) In-vitro evaluation of selected Egyptian traditional herbal medicines for treatment of alzheimer disease. BMC Complement Altern Med 13:121

Shrivastava N, Srivastava A, Banerjee A, Nivsarkar M (2006) Anti-ulcer activity of *Adhatoda vasica* Nees. J Herb Pharmacother 6(2):43–49

Singh TP, Singh OM, Singh HB (2011) *Adhatoda vasica* nees: phytochemical and pharmacological profile. J Nat Prod 1:29–39

The Wealth of India (2003) NISCAIR, Council of Scientific and Industrial Research. New Delhi Vol 1: pp 76–79

Uniyal SK, Singh KN, Jamwal P, Lal B (2006) Traditional use of medicinal plants among the tribal communities of Chhota Bhangal, Western Himalaya. J Ethnobiol Ethnomed 2(1):14

Yadav AK, Tangpu V (2008) Anticestodal activity of *Adhatoda vasica* extract against Hymenolepis diminuta infections in rats. J Ethnopharmacol 119:322–324

Chapter 4
Ethnomedicinal Facts and Practice

Ethnomedicine is the interaction between human beings and medicine which is an ancient aspect of practice to treat the ailments. Initially it was based on hit and trial methods, but the technological advancements opened the many ways to be more specific for the practice and outcome. It was only the verbal communication in the past to convey the information about the drugs of natural origins from generation to generation. Various systems (Ayurveda, Unani, Siddha and homeopathy) are known to have their practice on plant/natural origin-based medicines where extract, decoction, ashes and tinctures are prepared and given (Dubey et al. 2004; Mukherjee and Wahile 2006). Urbanization in modern era has affected the practice, but its importance can never be ruled out. Study of natural drugs is incomplete without taking ethnobotanical aspects into account to link it with ethnomedicinal aspects (Singh and Huidrom 2013).

Adhatoda vasica plant as being the promising and very dynamic indigenous to Asian region holds the attractive medicinal aspects and practice. The plant is in use since the ancient times and documented in various literatures. A comprehensive review pertaining to Southeast Asia was published and suggested the potential of the plant to be worked upon (Murali et al. 2018) by the scientific research scholars. The pharmacological aspects are mentioned in most elaborated and eloquent form in a separate section of the current book. The plant comprises the historical importance to be utilized as remedy when the techniques were not as sound as we observe in modern times. The people were bound to use it as traditional or folklore practice. Those practices are reported by various scholars for the ease of future explorations. Those all aspects are comprehensively summarized in Table 4.1, which includes the major categories of ailments along with the instructions to use it with different modes of preparations.

Table 4.1 Ethnomedicinal aspects of *A. vasica* for various ailments

Ailment categories	Mode of formulation/ preparation	Usage instruction	References
Respiratory diseases (cough, cold, bronchitis, tuberculosis and pneumonia) and malarial fever	Decoction of whole plant	**Asthma** – 5 ml decoction of mature leaves, two times in a day. Root decoction also. Root bark decoction with honey	Dulla and Jahan (2017), Haq (2012), Kumari et al. (2013), Poonam and Singh (2009), Revathi and Parimelazhagan (2010), Bhowmik et al. (2013), Goswami et al. (2013), Kaur and Kaur (2017), Mannaf et al. (2013), Shadangi et al. (2012), Sen et al. (2011), Shiddamallayya et al. (2010), Singh et al. 2010; Uniyal et al. (2002), Khan and Singh (2010), Sahani and Mall (2013), Rai and Lalramnghinglova (2010)
		Tuberculosis – leaf juice with goat milk for 6 months	
		Cough – decoction of the grounded powder of the following is given twice a day [15 leaves of vasaka + 15 leaves *Tylophora indica* + handful of *Albizia amara* + 1 or 2 leaves of *Aloe barbadensis* + *Piper nigrum* 10 seeds + 1 *Allium sativum* + Cherry (100 g)]. Root decoction also, leaf decoction with honey – three times a day. 5 leaves sprinkled with one pinch of rock salt and boiling with water.	
		Bronchitis – leaves and root decoction. Root decoction 10 ml for 7 days.	
		Malarial fever – root and leaves with ginger. 5 leaves sprinkled with one pinch of rock salt and boiling with water, use twice a day. 20 ml decoction of boiled leaves, two times a day for 3 days	

(continued)

Table 4.1 (continued)

Ailment categories	Mode of formulation/ preparation	Usage instruction	References
	Extract of different parts	**Asthma** – extract of flower along with *Solanum surattense*. Leaf extract with honey and jiggery, twice in a day for 3–5 days. Leaf and bark juice. Leaf and flower juice with *Hibiscus rosa-sinensis*. Leaf and root extract. Leaf extract with *Terminalia arjuna* bark and *Helicteres isora*, twice a day for 1 week.	Rauf et al. (2012), Das et al. (2012),Deepa and Saravanakumar (2013), Murthy and Vidyasagar (2013), Padal and Viyayakumar (2013), Das et al. (2008), Rahman et al. (2010), Tuhin et al. (2013), Muthu et al. (2006), Rai and Lalramnghinglova (2010), Rashid et al. (2013), Sadale and Karadge (2013), Sarmah et al. (2008), Hazrat et al. 2011; Rahmatullah et al. (2009b), Panda (2010), Hossain and Hoq (2016), Masum et al. (2013), Rahmatullah et al. (2009a), Bhat and Negi (2006), dahare and Jain (2010), Shende (2017)
		Whooping cough – leaf juice.	
		Bronchitis – leaf and root extract.	
		Cough – few drops of leaf extract orally for kids. Leaf extract with sugar three times a day for 1 week. Leaf and bark juice. Five leaves with honey.	
		Fever – leaf and root extract.	
		Phlegm – leaf and bark juice.	
		Pneumonia – root extract. Leaf extract.	
		Bleeding nose – leaf juice	
	Single or mixed paste of the different parts	**Chronic malaria** – leaf paste all over the body.	Asharaf and Sundaramari (2017), Hussain and Hore (2007), Savithramma et al. (2007)
		Asthma – mixed paste of *A. vasica* and roots of *Solanum surattense* along with fruit *Piper longum* (equal proportion). With honey for 1 week	
	Inhalational therapy	**Asthma and chronic bronchitis** – dry leaf powder and smoke from burning dry leaves	Desale et al. (2013), Ningthoujam et al. (2013)

(continued)

Table 4.1 (continued)

Ailment categories	Mode of formulation/ preparation	Usage instruction	References
	Dried powder of different parts	**Cough tuberculosis and asthma** – leaf + seeds of *Foeniculum vulgare* + rhizome of *Zingiber officinale* and *Terminalia chebula* thrice a day for 8–10 days. Root powder only for asthma. Bark powder for pulmonary effusion and asthma. Leaf powder with spoonful honey for cough and coryza.	Abbasi et al. (2010), Poonam and Singh (2009), Sahani and mall (2013)
		Malaria – leaf powder	
Diabetes	Extracts and juices of different parts	**Diabetes** – extract of leaves of *A. vasica* + *Clerodendrum indicum* and/or *Azadirachta indica* + *Zanthoxylum acanthopodium*.	Khan and Yadava (2010), Mootoosamy and Mahomoodally (2014), Mannaf et al. (2013), Sahani and Mall (2013), Ahmad et al. (2004), Rauf et al. (2012)
		Juices of leaves of *A. vasica* + *Andrographis paniculata* for 21 days. Chewing of young fresh leaves (2) at empty stomach. Extract of flower + *Solanum surattense*. Flowers + *Azadirachta indica* + Gum of *Acacia nilotica*	
Abdominal pain, other pains and nausea	Decoction	**Ear pain** – leaf decoction as antispasmodic to relieve ear pain	Hazrat et al. (2011)
	Warmed up leaves	**Lumber pains and sprains** – few leaves are warmed up on fire and plastered over the joints or surface required	Jamir et al. (1999)
	Extracts	**Stomach pain** – extract of stem and root bark	Rajith et al. (2010), Haq (2012), Rahman et al. (2010), Shiddamallayya et al. (2010)
		Antispasmodic – leaf and root extract	
		Muscular spasm – flower and fruit extract	
		Post-delivery bath – fresh leaves extract with other drugs extracts, i.e. mature leaves of *Careya arborea*, *Calycopteris floribunda*, *Quassia indica*, *Clerodendrum infortunatum*, *Musa paradisiaca*, *Tamarindus indica*	
		Antiemetic – leaf and bark juice	
	Leaf paste	**Sprains** – leaf paste applied	Hussain and Hore (2007)
		Ear pain and headache – leaf paste to be taken orally	
Gastric trouble	Extract	Leaf extract is given	Sonowal and Barua (2011)

(continued)

Table 4.1 (continued)

Ailment categories	Mode of formulation/ preparation	Usage instruction	References
Anthelmintic	Extract	Root extracts are used to expel out and young leaf and bark juice for the mentioned purpose as well	Rahman et al. (2010), Rai and Lalramnghinglova (2010)
Dysentery and diarrhoea	Extracts	**Anti-diarrhoeal** – leaf extract (one tea spoonful) twice a day. **Dysentery** – fresh root extract	Sarmah et al. (2008), Sen et al. (2011), Bhatt and Negi (2006), Hussain and Hore (2007), Jamir and Takatemjen (2010), Rai and Lalramnghinglova (2010), Sarmah et al. (2008), Basumatary et al. (2004)
	Decoction	**Anti-dysentery** – leaf decoction and juice	Hazrat et al. (2011), Sen et al. (2011), Shanmugam et al. (2011)
Liver disorders	Extract and decoction	**Jaundice** – root , flower and leaf extract, two spoonful extract with sugar, leaf and stem decoction with honey, whole plant extract for liver fever	Singh et al. (2010), Sadale and Karadge (2013), Rahim et al. (2012), Jamir and Takatemjen (2010), Rauf et al. (2012)
Arthritis, inflammation, rheumatism, cuts and wounds	Extracts and juices	**Arthritis**- 2 tea spoonful leaf extract. **Rheumatism** – root extract and decoction applied over surface. **Pimples** – floral extract with mustard oil. **Cuts and wounds** – leaf juices applied externally	Dutta (2017), Srivastava and Samuel (2013), Jamir and Takatemjen (2010), Rauf et al. (2012)
	Paste and poultice	**Gout** – leaf paste is applied **Cuts and wounds** – leaf paste	Hussain and Hore (2007), Rai and Lalramnghinglova (2010)
Ophthalmic and antiseptic	Extract and decoction	**Ophthalmic use** – fresh flower extract **Antiseptic** – root extract and leaf decoction	Masum et al. (2013), Rai and Lalramnghinglova (2010), Hazrat et al. (2011)
Fits	Decoction and juices	**Epilepsy** – juices of leaves of *A. vasica + Zingiber officinalis + Piper nigrum* and beetle leaf	Alagesaboopathi (2011)

(continued)

Table 4.1 (continued)

Ailment categories	Mode of formulation/ preparation	Usage instruction	References
Thrombopoietic	Paste	**Blood clotting** – leaf paste	Rajan et al. (2002)
Paralysis	Decoction	**Cattle paralysis** – decoction of roots + bark of *Oroxylum indicum*+ bark of *Terminalia paniculata* + bark of *Trachyspermum ammi* + *Piper nigrum*	Hussain and Hore (2007), Shadangi et al. (2012), Rai and Lalramnghinglova (2010)
Gynaecological problems	Juices and decoction	**Leucorrhoea** – root bark juice with honey.	
		Menstrual irregularities – few leaves + 1 g *Daucus carota* + *Raphnus sativus* seeds	
		Labour pain – root paste applied over abdomen	
		Gonorrhoea – roots and leaves decoction with ginger	
Toxic bites	Leaf paste and juice	**Scorpion sting** – leaf paste and juice applied over surface	Rao et al. (2006)
	Decoction	**Snake bite** – root and leaf decoction of *A. vasica* + *Alangium salvifolium* + *Coccinia grandis* given orally	Ayyanar and Ignacimuthu (2005)

References

Abbasi AM, Khan MA, Ahmad M, Zafar M (2010) Herbal medicines used to cure various ailments by the inhabitants of Abbottabad district, north west Frontier Province, Pakistan. Indian J Tradit Knowl 9(1):175–183

Ahmad M, Khan MA, Arshad M, Zafar M (2004) Ethnophytotherapical approaches for the treatment of diabetes by the local inhabitants of district Attock (Pakistan). Ethnobot Leaf 1:1–10

Alagesaboopathi C (2011) Ethnomedicinal plants used as medicine by the Kurumba tribals in Pennagaram region, Dharmapuri District of Tamil Nadu, India. Asian J Exp Biol Sci 2(1):140–142

Asharaf S, Sundaramari M (2017) A quantitative study on indigenous medicinal plants used by tribes of Kerala. J Ext Educ 28(3):5695–5702

Ayyanar M, Ignacimuthu S (2005) Medicinal plants used by the tribals of Tirunelveli hills, Tamil Nadu to treat poisonous bites and skin diseases. Indian J Tradit Knowl 4(3):229–236

Basumatary SK, Ahmed M, Deka SP (2004) Some medicinal plant leaves used by Boro (tribal) people of Goalpara district, Assam. Nat Prod Radiance 3(2):88–90

Bhatt VP, Negi GC (2006) Ethnomedicinal plant resources of Jaunsari tribe of Garhwal Himalaya, Uttaranchal. Indian J Tradit Know 5(3):331–335

Bhowmik S, Datta BK, Sarbadhikary SB, Mandal NC (2013) Contribution to the less known ethnomedicinal plants used by Munda and Santal community of India with their ethnomedicinal justification. World Appl Sci J 23(10):1408–1417

Dahare DK, Jain A (2010) Ethnobotanical studies on plant resources of Tahsil Multai, district Betul, Madhya Pradesh, India. Ethnobot Leaf 14:694–705

Das AK, Dutta BK, Sharma GD (2008) Medicinal plants used by different tribes of Cachar district, Assam. Indian J Tradit Knowl 7(3):446–454

Das PR, Islam MT, Mahmud ASMSB, Kabir MH, Hasan ME, Khatun Z, Rahman MM, Nurunnabi M, Khatun Z, Lee YK et al (2012) An ethnomedicinal survey conducted among the folk medicinal practitioners of three villages in Kurigram district, Bangladesh. Am Eurasian J Sustain Agric 6(2):85–96

Deepa J, Saravanakumar K (2013) Traditional phytotheraphy in Chidambaram taluk of Cuddalore district, Tamil Nadu. Curr Res Med Med Sci 3(1):1–5

Desale MK, Bhamare PB, Sawant PS, Patil SR, Kamble SY (2013) Medicinal plants used by the rural people of Taluka Purandhar, District Pune, Maharashtra. Indian J Traditional Knowl 12(2):334–338

Dubey NK, Kumar R, Tripathi P (2004) Global promotion of herbal medicine: India's opportunity. Curr Sci 86(1):37–41

Dulla O, Jahan FI (2017) Ethnopharmacological survey on traditional medicinal plants at Kalaroa Upazila, Satkhira district, Khulna division, Bangladesh. J Int Ethnopharmacol 6(3):316

Dutta ML (2017) Plants used as ethnomedicine by the Thengal Kacharies of Assam, India. Asian J Plant Sci Res 7(1):7–8

Goswami H, Hassan MR, Rahman H, Islam E, Asaduzzaman M, Prottoy MA, Seraj S, Rahmatullah M (2013) Ethnomedicinal wisdom of the Tripura tribe of Comilla district. A combination of medicinal plant knowledge and folk beliefs. American, Bangladesh

Haq F (2012) The ethnobotanical uses of medicinal plants of Allai Valley, Western Himalaya, Pakistan. Int J Plant Res 2(1):21–34

Hazrat A, Nisar M, Shah J, Ahmad S (2011) Ethnobotanical study of some elite plants belonging to Dir, Kohistan valley, Khyber Pukhtunkhwa, Pakistan. Pak J Bot 43(2):787–795

Hossain MT, Hoq MO (2016) Therapeutic use of *Adhatoda vasica*. Asian J Med Biol Res 2(2):156–163

Hussain S, Hore DK (2007) Collection and conservation of major medicinal pants of Darjeeling and Sikkim Himalayas. Indian J Traditional Knowl 6(2):352–357

Jamir TT, Sharma HK, Dolui AK (1999) Folklore medicinal plants of Nagaland, India. Fitoterapia 70(4):395–401

Jamir NS, Takatemjen L (2010) Traditional knowledge of Lotha-Naga tribes in Wokha District, Nagaland. Indian J Tradit Knowl 9(1):45–48

Kaur R, Kaur H (2017) Plant derived antimalarial agents. J Med Plants 5(1):346–363

Khan JB, Singh GP (2010) Ethno-medicinal active plants for treating cold and cough in the vicinity of Nahargarh Wildlife Sanctuary, Jaipur, India. Our Nat 8(1):225–230

Khan MH, Yadava PS (2010) Antidiabetic plants used in Thoubal district of Manipur, Northeast India. Indian J Tradit Knowl 9(3):510–514

Kumari S, Batish DR, Singh HP, Negi K, Kohli RK (2013) An ethnobotanical survey of medicinal plants used by Gujjar community of Trikuta Hills in Jammu and Kashmir, India. J Med Plant Res 7(28):2111–2121

Mannaf MA, Islam MA, Akter S, Akter R, Nasrin T, Zarin I, Seraj S, Rahmatullah M (2013) A randomized survey of differences in medicinal plant selection as well as diseases treated among folk medicinal practitioners and between folk and tribal medicinal practitioners in Bangladesh. American-Eurasian J Sustain Agricult 7(3):196–209

Masum GZH, Dash BK, Barman SK, Sen MK (2013) A comprehensive ethnomedicinal documentation of medicinal plants of Islamic university region, Bangladesh. Int J Pharma Sci Res 4(3):1202–1209

Mootoosamy A, Mahomoodally MF (2014) Ethnomedicinal application of native remedies used against diabetes and related complications in Mauritius. J Ethnopharmacol 151(1):413–444

Mukherjee PK, Wahile A (2006) Integrated approaches towards drug development from Ayurveda and other Indian system of medicines. J Ethnopharmacol 103(1):25–35

Murali RS, Rao GN, Basavaraju R (2018) Ethnomedicinal importance of *Adhatoda vasica* in the South East Asian countries: review and perspectives. Stud Ethno Med 12(2):120–131

Murthy SMS, Vidyasagar GM (2013) Traditional knowledge on medicinal plants used in the treatment of respiratory disorders in Bellary district, Karnataka, India. Indian J Nat Prod Resour 4(2):189–193

Muthu C, Ayyanar M, Raja N, Ignacimuthu S (2006) Medicinal plants used by traditional healers in Kancheepuram District of Tamil Nadu, India. J Ethnobiol Ethnomed 2(1):43

Ningthoujam SS, Talukdar AD, Potsangbam KS, Choudhury MD (2013) Traditional uses of herbal vapour therapy in Manipur, North East India: an ethnobotanical survey. J Ethnopharmacol 147:136–147

Padal SB, Viyayakumar Y (2013) Traditional knowledge of Valmiki tribes of G. Madugula Mandalam, Visakhapatnam district, Andhra Pradesh. Int J Innov Res Dev 2(6):723–738

Panda T (2010) Preliminary study of ethno-medicinal plants used to cure different diseases in coastal district of Orissa, India. Br J Pharmacol Toxicol 1(2):67–71

Poonam K, Singh GS (2009) Ethnobotanical study of medicinal plants used by the Taungya community in Terai Arc landscape, India. J Ethnopharmacol 123(1):1167–1176

Rahim ZB, Rahman MM, Saha D, Hosen SMZ, Paul S, Kader S (2012) Ethnomedicinal plants used against jaundice in Bangladesh and its economical prospects. Bull Pharma Res 2(2):91–105

Rahman AHMM, Kabir EZM, Sima SN, Sultana RS, Nasiruddin M, Zaman ATMN (2010) Study of an ethnobotany at the village Dohanagar, Naogaon. J Appl Sci Res 6(9):1466–1473

Rahmatullah M, Das AK, Mollik ARH, Jahan R, Khan M, Rahman T, Chowdhury MH (2009a) An ethnomedicinal survey of Dhamrai Sub-District in Dhaka District, Bangladesh. Am Eurasian J Sustain Agric 3(4):881–888

Rahmatullah M, Mukti IJ, Haque AKMF, Mollik MAH, Parvin K, Jahan R, Chowdhury MH, Rahman T (2009b) An ethnobotanical survey and pharmacological evaluation of medicinal plants used by the Garo tribal community living in Netrakona district, Bangladesh. Adv Nat Appl Sci 3(3):402–418

Rai PK, Lalramnghinglova H (2010) Ethnomedicinal plant resources of Mizoram, India: implication of traditional knowledge in health care system. Ethnobot Leaf 14:274–305

Rajan S, Sethuraman M, Mukherjee PK (2002) Ethnobiology of the Nilgiri hills, India. Phytother Res 16(2):98–116

Rajith NP, Navas M, Thaha AM, Manju MJ, Anish N, Rajasekharan S, George V (2010) A study on traditional mother care plants of rural communities of South Kerala. Indian J Tradit Knowl 9(1):203–208

Rao DM, Bhaskara R, Gudivada S (2006) Ethno-medico-botanical studies from ayalaseema region of southern Eastern Ghats, Andhra Pradesh, India. Ethnobot Leaf 10:198–207

Rashid A, Tariq SR, Chowdhury ZZ, Rashid SA, El Sherbini AM, Al-Fedaghi S (2013) Ethnomedicinal plants used in the traditional phytotherapy of chest diseases by the Gujjar-Bakerwal tribe of district Rajouri of Jammu and Kashmir state. Int J Pharma Sci Res 4(1):328–333

Rauf F, Qureshi R, Shaheen H (2012) Folk medicinal uses of indigenous plant species of Barroha, Bhara- Kahu and Maanga in Islamabad, Pakistan. J Med Plant Res 6(11):2061–2070

Revathi P, Parimelazhagan T (2010) Traditional knowledge on medicinal plants used by the Irula tribe of Hasanur hills, Erode district, Tamil Nadu, India. Ethnobot Leaf 14:136–160

Sadale AN, Karadge BA (2013) Survey on ethnomedicinal plants of Ajara Tahsil, District Kolhapur, Maharashtra (India). Trends Life Sci 2(1):21–25

Sahani S, Mall TP (2013) Ethnomedicinal plants from Bahraich (UP) India. Indian J Sci 2(5):112–120

Sarmah R, Adhikari D, Mazumder M, Arunachalam A (2008) Traditional medicobotany of Chakma community residing in the northwestern periphery of Namdapha National Park in Arunachal Pradesh. Indian J Tradit Knowl 7:587–593

Savithramma N, Sulochana CH, Rao KN (2007) Ethnobotanical survey of plants used to treat asthma in Andhra Pradesh, India. J Ethnopharmacol 113(1):54–61

Sen S, Chakraborty R, De B, Devanna N (2011) An ethnobotanical survey of medicinal plants used by ethnic people in West and South Districts of Tripura, India. J Forestry Res 22(3):417–426

Shadangi AK, Panda RP, Patra AK (2012) Ethnobotanical studies of wild flora at G. Udayagiri forest in Eastern Ghats, Odisha. J Environ Sci Toxicol Food Technol 2(2):25–37

Shanmugam S, Annadurai M, Rajendran K (2011) Ethnomedicinal plants used to cure diarrhea and dysentery in Pachalur hills of Dindigul district in Tamil Nadu, Southern India. J Appl Pharma Sci 1(8):94–97

Shende CB (2017) Ethnomedicinal study of Gadchiroli reserve forests Maharashtra. Int J Appl Res 3(3):100–102

Shiddamallayya N, Yasmeen A, Gopakumar K (2010) Hundred common forest medicinal plants of Karnataka in primary healthcare. Indian J Tradit Knowl 9(1):90–95

Singh KJ, Huidrom D (2013) Ethnobotanical uses of medicinal plant, Justicia adhatoda L. by Meitei community of Manipur, India. J Coast Life Med 1(4):322–325

Singh PK, Kumar V, Tiwari RK, Sharma A, Rao CV, Singh RH (2010) Medico-ethnobotany of Chatara block of District Sonebhadra, Uttar Pradesh, India. Adv Biol Res 4(1):65–80

Sonowal R, Barua I (2011) Ethnomedical practices among the Tai-Khamyangs of Assam, India. Ethnomed 5(1):41–50

Srivastava S, Samuel CO (2013) Herbal cures practised by rural populace in Varanasi region of eastern UP (India). IOSR J Pharma Biol Sci 6(1):1–5

Tuhin MIH, Asaduzzaman M, Islam E, Khatun Z, Rahmatullah M (2013) Medicinal plants used by folk medicinal herbalists in seven villages of Bhola District, Bangladesh. Am Eurasian J Sustain Agric 7(3):210–218

Uniyal SK, Awasthi A, Rawat GS (2002) Traditional and ethnobotanical uses of plants in Bhagirathi valley (Western Himalayas). Indian J Tradit Knowl 1(1):7–19

Chapter 5
Biotechnological Approaches and Production of Secondary Metabolites

5.1 Approach and Initiatives

A regular immoderate exploitation of the significant plant species having the less opportunity for natural propagation put them in endangered plant categories which lead to have an urge to develop the techniques to meet the requirement of their products. It is our daily requirement we cannot refrain in one or other ways. Our remedies to cure and alleviate the ailments have direct or indirect relationships to the natural sources. *A. vasica* as the promising herbal remedy to cure the respiratory disorders is the attribute of therapeutic response that appears through quinazoline alkaloids. These alkaloids have many biogenetic pathways to get synthesized in the different parts of the plant. Studies at molecular level suggest us to adopt the protocols to enhance these alkaloidal metabolites which include the tissue culture studies and employment of bio elicitors. The utilities and propagation is bound to grow the cell contents under controlled conditions which act later like the primary source to provide the phytochemicals to develop the different formulations after direct and indirect organogenesis (Bhambhani et al. 2012; Abhyankar and Reddy 2007; Murali and Basavaraju 2012; Panigrahi et al. 2017). The purpose remains intact to produce the more amounts of secondary metabolites by preserving the explant source and seed germplasm to conquer over the scarcity of natural availabilities. The rate limiting factors also include an unavoidable poor rate of seed germination of *A. vasica* for normal clonal propagation (Mathew et al. 1998; Bhambhani et al. 2012), hence declined phytochemical productions. The application of these approaches through plant biotechnology for the purpose of clonal propagation of phyto-medicinals has the capacity to meet the requirement of raw materials with great compliance (Pierik 1987).

M. Ali, K. R. Hakeem, *Scientific Explorations of Adhatoda vasica*,
SpringerBriefs in Plant Science, https://doi.org/10.1007/978-3-030-56715-6_5

The results of in vitro regeneration under controlled conditions in tissue culture techniques reveals the most preferred time to collect the explant from *A. vasica*, between November and March (Abhyankar and Reddy 2007; Mandal and Laxminarayana 2014). This timing and its suitability indicates the maximum cell viability to proliferate in artificial media. The probability of contamination remains high because of soil microbes and environmental pollutants. It requires proper sterilization to prevent any microbial growth.

Surface sterilization of different explants for regeneration response in *A. vasica* was correlated and summarized with great efforts by Gantait and Panigrahi in 2018, which might be helpful to design a successful protocol for *A. vasica* explants in the future. Explants in vitro culture studies are bound to go through many challenges for the final establishment where microbial attack is one among due to the most vulnerable nature (Gantait et al. 2014). Surface sterilization and method development is required to be standardized by taking three most important parameters into consideration. Parameters include the type of disinfectant, its concentration and time of exposure to the explants. The standardization of this protocol assures the researchers that the regeneration ability of the different explants from *A. vasica* would be uninterrupted despite eradicating the contaminants only. Different explants of *A. vasica* plant (i.e. leaf, nodal segments, root and shoot tips, stem and petiole) require a variable concentrations and exposure durations. Generally the younger juvenile explants need simple disinfectants with minimum concentrations for less exposure timings in comparison to the harder and mature tissues (Gantait and Kundu 2017; Jaiswal et al. 2017).

Huge several experimentation and trials reveal the fact of effectiveness of the sterilization protocols. A minute change in the concentration and time of exposure of the disinfectant directly affects the response of regenerations in newly developing cells. Concentration ranges and time of exposure applied on various *A. vasica* explants are summarized (Gantait and Panigrahi 2018) in Table 5.1.

Surface sterilization as being the most crucial step in biotransformation and regeneration process, given in Table 5.1, suggests mercuric chloride concentration range from 0.01% to 0.2% w/v with its exposure timings from 1 to 18 min. Timing of exposure or contact variance is applicable to all the explants, but it depends on the nature and type of explants. Other disinfectants, i.e. Savlon, ethanol, Dettol, Bavistin solution and hydrogen peroxide, also follow the same pattern of usage concept to keep the explants free from any microbial contaminations.

Proper surface sterilization prepares the explant to be processed for in vitro regeneration via the standardization of basal nutrient media, carbohydrates, vitamins, solidifying agents (i.e. agar), adjusted pH and the calculated concentration of growth regulators for the phases of micro-propagation of vasaka explant. Most preferred basal media for *A. vasica* has been reported by various journals possessing the claim of result reproducibility is Murashige and Skoog (MS) media (Murashige and Skoog 1962).

Table 5.1 Disinfectants with concentration and exposure timings

S. no.	Disinfectant	Concentration range	Time of exposure	Additional remark
1.	Mercuric chloride ($HgCl_2$)	0.01–0.2% w/v	Not less than 1 min to not more than 18 min	Depends on the type of explant. The time exposure is conditional
2.	Savlon	1% w/v	Not less than 10 min to not more than 20 min	Depends on the type of explant. The time exposure is conditional
3.	Ethanol	70–95% v/v	Not less than 30 s to not more than 1 min	Depends on the type of explant. The time exposure is conditional
4.	Dettol	1%	Approximately 10 min	Depends on the type of explant. The time exposure is conditional
5.	Bavistin solution	1% w/v	Approximately 10 min	Depends on the type of explant. The time exposure is conditional
6.	Hydrogen peroxide (H_2O_2)	3%	2 min	Depends on the type of explant. The time exposure is conditional

Surfactants Teepol and Tween 20 and 80 would be employed in very less concentrations and rinsing would always be carried out with sterilized water

5.2 Shoot, Callus Initiation and Regeneration

Shoot multiplication and its enhancement remains the objectives of the researchers to have more yield and productivity. A list of most preferred plant growth regulators allows the researchers to use as a sole regulator or in combinations. To standardize the method is quite cumbersome at the initial stages, but the adequate attention by the researchers made it easy to standardize with the suggestive combinations of growth regulators. The reports suggest the combination of cytokinin and auxin as 0.5–2 mg/l N^6-benzyladenine (BA) and 0.05–0.2 mg/l α-naphthalene acetic acid (NAA), respectively (Murali and Basavaraju 2012; Azad et al. 2003; Khalekuzzaman et al. 2008; Amin et al. 1997).

1 mg/l kinetin (6-furfurylaminopurine) is the source of cytokinin which is responsible for enhanced shoot multiplication in the combination of equal concentration of N^6-benzyladenine (BA) (Abhyankar and Reddy 2007), while Roja et al. (2011) suggested a combination of gibberellin A3 (GA3) 1 mg/l + BA 1 mg/l to achieve the same enhanced shoot multiplication. The composition of 0.5 mg/l α-naphthalene acetic acid (NAA) and 0.5 mg/l thidiazuron (TDZ) with 2 mg/l N^6-benzyladenine (BA) was reported by Lone et al. to make BA more efficient for regeneration. Maximum shoot production is reported in 28 days almost in all the explants.

Callus induction was always preferred with MS basal media of choice but Anand and Bansal suggested (Anand and Bansal 2002) and chose Gamborg's medium (B5) (Gamborg et al. 1968) basal media to induce the callus by using the leaf explants with subsequent supplementation of additional 1 mg/l 2,4 -D.

Direct organogenesis of *A. vasica* is dependent upon the category and concentrations of the plant growth regulators (PGRs). Equal or variable proportions of auxin and cytokinin have been observed to induce a significant frequency of friable calli or organogenic calli. Callus induction in terms of percentage was calculated by the researchers with variable days after inoculation. Mandal and Laxminarayana obtained callus induction 100% in MS medium with subsequent supplementation of 0.25 mg/l NAA and TDZ (Mandal and Laxminarayana 2014), while Dinesh and Parameswaran mentioned 90% callus induction within 7 days of inoculation with higher ratio of PGRs (Dinesh and Parameswaran 2009). Trend analogy was found by Bhambhani et al. and reported 70% callus induction in leaf explants within 4 weeks of inoculation steps (Bhambhani et al. 2012), while Singh and Sharma obtained highly considerable frequency of friable calli in NAA and BA combination (Singh and Sharma 2014).

Changing explant might be an approach to have the better results for callus induction. Sil and Ghosh observed the opposite trends by using nodal segments as explant from *A. vasica* (Sil and Ghosh 2010) supplemented with higher ratio of cytokinin and auxin. Dual combination of auxin/cytokinin showed the interesting results (75% callus along with 18.16 g fresh weight) in a study performed by Maurya and Singh. It was proved a very novel approach of amalgamation of two sets of PGRs by supplementation to MS medium (Maurya and Singh 2010). Single use of sole auxin (1 mg/l 2,4 – D) was reported by Panigrahi et al. to induce 46% of calli through nodal segment as explant of *A. vasica* without adventitious shoots (Panigrahi et al. 2017). These in vitro studies also suggest that the root can be obtained from callus by suppressing the shoot regeneration. Jayapaul et al. had reported earlier of somatic embryogenesis after 62% of callus induction with 4.5 μM 2,4-D & 2.3 μM Kn combination supplemented to MS medium (Jayapaul et al. 2005). Though studies on *A. vasica* to reveal somatic embryogenesis is the segment of future perspectives.

Various combinations of PGRs supplement along with basal medias have been attempted to find all the possibilities of regeneration of shoot root and calli. The summarized forms of the combinations and concentrations of plant growth regulators are listed in Table 5.2; it comprises all the approaches as the guidance for future perspectives especially for *A. vasica* explant.

5.3   Seed Production: Biotechnological Approach

Artificial seed production technology is useful for the precious endangered plant species to preserve the germplasm without affecting the natural population. It is multifaceted and dynamic in itself and very propitious to the research scholars

Table 5.2 Plant growth regulators and their strengths for different explants of *A. vasica*

S. no.	Explants of *A. vasica*	Media and supplemented PGRs	References
1.	Leaf	MS + 1.0 mg/l BA and 0.1 mg/l NAA	Amin et al. (1997)
		B5 + 1 mg/l 2,4-D	Anand and Bansal (2004)
		B5 + 0.1 mg/l Kn	
		MS + 21.5 μM NAA + 19.7 M μBA and 9.3 μM Kn	Jayapaul et al. (2005)
		MS + 10.7 μM NAA and 2.2 μM BA	Dinesh and Parameswaran (2009)
		MS + 1.5 ppm 2,4-D + 1.5 ppm IAA + 1.5 ppm Kn and 1.5 ppm BA	Maurya and Singh (2010)
		MS + 1 mg/l 2,4-D and 0.5 mg/l Kn	Bhambhani et al. (2012)
		MS + 6 mg/l IAA and 6 mg/l Kn	Rashmi et al. (2012)
		MS + 1 ppm 2,4 D + 1 ppm BA and 1 ppm IAA	Madhukar et al. (2014)
2.	Nodal segment	MS + 0.5 mg/l BA and 0.1 mg/l NAA	Azad et al. (2003)
		MS + 10 mg/l BA	Abhyankar and Reddy (2007)
		MS + 2.0 mg/l BA and 0.5 mg/l NAA	Sil and Ghosh (2010)
		MS + 15% (v/v) CW and 5 mg/l BA	Bimal and Shahnawaz (2012)
		MS + 1.0 mg/l BA and 0.05 mg/l NAA	Murali and Basavaraju (2012)
		MS + 3 mg/l IBA and 3 mg/l BA	Rashmi et al. (2012)
		MS + 10.0 mg/l BA	Khan et al. (2016)
		MS + 1.1 mg/l BA	Panigrahi et al. (2017)
3.	Shoot tip	MS + 0.5 mg/l BA and 15% CM	Nath and Buragohain (2005)
		MS + 2 mg/l BA and 0.2 mg/l NAA	Khalekuzzaman et al. (2008)
		MS + 22.20 μm BA	Tejavathi et al. (2008)
		MS + 2 mg/l BA +0.5 mg/l NAA and 0.5 mg/l TDZ	Lone et al. (2013)
4.	Petiole	MS + 4.5 μM 2,4-D and 2.3 μM Kn	Jayapaul et al. (2005)
		MS + 0.25 mg/l TDZ and 0.25 mg/l NAA	Mandal and Laxminarayana (2014)
5.	Root	MS + 3 mg/l IBA and 6 mg/l BA	Rashmi et al. (2012)
		MS + 3.5 mg/l NAA and 1.25 mg/l BA	Singh and Sharma (2014)
		MS + 1 mg/l 2,4-D and 4 mg/l BA	Singh et al. (2017)

Abbreviations

2,4-D 2,4-dichlorophenoxyacetic acid, *B5* medium, or Gamborg's medium, BA N 6-benzyladenine, *CM* coconut milk, *CW* coconut water, *IAA* indole-3-acetic acid, *IBA* indole-3-butyric acid, *Kn* kinetin or 6-furfuryl aminopurine, *MS* Murashige and Skoog medium, *NAA* α-naphthalene acetic acid, *TDZ* thidiazuron

willing in vitro regeneration and propagation studies. It helps to conserve the plant with its genetic identity (Gantait et al. 2015; Haque and Ghosh 2016). Synthetic seeds comprise two essential components which are plant propagule (might be in vitro or in vivo derived) and secondly the matrix (a gelling material which is required to encapsulate plant propagules). Inclusion of nutrients and antibiotics with other essential additives completes a protocol for the purpose (Sharma et al. 2019), although it has not been much reported or done about *A. vesica* to develop an artificial seed. Rare references are there in public domain to document it. Anand and Bansal developed and solely reported about the artificial seed of *A. vasica*. They also reported about the highest conversion using synthetic seeds which were prepared in B5 medium having kinetin (Kn) and phloroglucinol (PG). However, declined response was also observed when repetitive growth medium was supplemented with kinetin (Kn) and phloroglucinol (PG) without taking encapsulation matrix into account (i.e. with or without growth adjuvants). The technology needs an explant (it might be a tissue, organ, or segment of the plant) to be poured into the solution of sodium alginate, and later on this aliquot having explant is dropped in another solution of calcium chloride ($CaCl_2$) to make the artificial seeds in spherical design. The technology also helps to store and exchange the plant samples for future prospective (Gantait et al. 2015; Gantait et al. 2017a, b). The flow diagram is attached as Fig. 5.1.

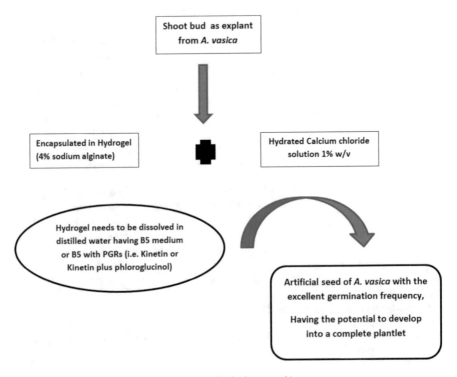

Fig. 5.1 Artificial seed production (biotechnological approach)

5.4 Secondary Metabolites in Response to Biotechnological Transformations

In vitro studies for organogenesis are more sustainable approaches to produce the secondary metabolites. *A. vasica* possess the medicinal components in all parts where vasicine and vasicinone as being the pyrroloquinazoline alkaloids dominate over the other categories. Jayapaul et al. published the presence of vasicine from callus culture of *A. vasica* showing the leaf explant is more probable to have more vasicine in culture studies especially when MS media is fortified with optimized plant growth regulators. Further chromatographic studies quantified the more significant presence of vasicine over vasicinone. The extracts from *A. vasica* in different solvents of changed polarities provide the varying concentrations of vasicine and vasicinone which can also be enhanced (as suggested by Bhambani et al.) through the biotechnological approaches by introducing the different elicitors. Elicitors comprise a range of compounds to be used in optimized ways. Chitosan, sodium salicylate, vitamin C, yeast extract and methyl jasmonate are the frequently used elicitors. Callus culture and suspension culture through the same protocols might be having the different yields corresponding to the dry weights. The reports regarding vasicinone yield are lesser than the vasicine which required more attention to be paid to consider this alkaloid more efficiently as well. Fungal endophytes are also traced histologically to check the symbiotic association with *A. vasica* which might facilitate the production of secondary metabolites. Localization and identification under controlled conditions are less time-consuming than the rest of the cultural studies, but obviously not free from the various confinements (Yash et al. 2015). These can be employed as bio elicitors.

References

Abhyankar G, Reddy VD (2007) Rapid micropropagation via axillary bud proliferation of *Adhatoda vasica* Nees from nodal segments. Indian J Exp Biol 45:268–271

Amin MN, Azad MAK, Begum F (1997) In vitro plant regeneration from leaf derived callus cultures of *Adhatoda vasica* Nees. Plant Tissue Cult 7:109–115

Anand Y, Bansal YK (2002) Synthetic seeds: a novel approach of in vitro plantlet formation in Vasaka (*Adhatoda vasica* Nees.). Plant Biotechnol 19(3):159–162

Anand Y, Bansal YKJ (2004) In vitro regeneration of *Adhatoda vasica* nees through adventitious organogenesis at diploil level of genome. Phytol Res 17:17–19

Azad MAK, Amin MN, Begum F (2003) Rapid clonal propagation of a medicinal plant-*Adhatoda vasica* nees. Using tissue culture technique. J Biol Sci 3(2):172–182

Bhambhani S, Karwasara VS, Dixit VK, Banerjee S (2012) Enhanced production of vasicine in *Adhatoda vasica* (L.)Nees.cell culture by elicitation. Acta Physiol Plant 34(4):1571–1578

Bimal R, Shahnawaz M (2012) Plant regeneration from nodal explants of *Adhatoda vasica* Nees. J Med Plants Res 6(7):1229–1233

Dinesh K, Parameswaran S (2009) Micropropagation and Organogenesis in *Adhatoda vasica* For The Estimation Of Vascine. Pharmacog Mag 5(20):359

Gamborg OL, Miller R, Ojima K (1968) Nutrient requirements of suspension cultures of soybean root cells. Exp Cell Res 50(1):151–158

Gantait S, Panigrahi J (2018) In vitro biotechnological advancements in Malabar nut (*Adhatoda vasica* Nees): achievements, status and prospects. J Genet Eng Biotechnol 16:545–552

Gantait S, Sinniah UR, Das PK (2014) *Aloe vera*: a review update on advancement of *in vitro* culture. Acta Agric Scand Sect B Soil Plant Sci 64(1):1–12

Gantait S, Kundu S, Ali N, Sahu NC (2015) Synthetic seed production of medicinal plants: a review on influence of explants, encapsulation agent and matrix. Acta Physiol Plant 37(5):98

Gantait S, Kundu S (2017) Neoteric trends in tissue culture-mediated biotechnology of Indian ipecac [Tylophora indica (Burm. f.) Merrill]. Biotech 7(3):231

Gantait S, Kundu S, Yeasmin L, Ali MN (2017a) Impact of differential levels of sodium alginate, calcium chloride and basal media on germination frequency of genetically true artificial seeds of Rauvolfia serpentina (L.) Benth. ex Kurz. J Appl Res Med Aromat Plants 4:75–81

Gantait S, Vijayan J, Majee A (2017b) Artificial seed production of Tylophora indica for interim storing and swapping of germplasm. Hortic Plant J 3(1):41–46

Haque SM, Ghosh B (2016) High-frequency somatic embryogenesis and artificial seeds for mass production of true-to-type plants in Ledebouria revoluta: an important cardioprotective plant. Plant Cell Tiss Org 127(1):71–83

Jaiswal N, Verma Y, Misra P (2017) Micropropagation and in vitro elicitation of licorice (Glycyrrhiza spp.). In Vitro Cell Dev Biol Plant 53(3):145–166

Jayapaul K, Kishor PK, Reddy KJ (2005) Production of pyrroloquinazoline alkaloid from leaf and petiole-derived callus cultures of Adhatoda zeylanica. J in vitro Cell Dev Biol Plant 41(5):682–685

Khalekuzzaman M, Rahman MS, Rashid M, Hossain MS (2008) High frequency in vitro propagation of *Adhatoda vasica* Nees through shoot tip and nodal explants culture. J Biol Sci 16:35–39

Khan S, Akter S, Habib A, Banu TA, Islam M, Khan NF, Islam S (2016) Establishment of in vitro regeneration protocol for *Adhatoda vasica* Nees. Bangladesh J Sci Ind Res 51(1):75–80

Lone SA, Yadav AS, Sharma AK, Tafazul M, Badkhane Y, Raghuwanshi DK (2013) A review on *Adhatoda vasica* Nees-an important and high demanded medicinal plant. Indo Am J Pharma Res 3:3341–3360

Madhukar G, Tamboli ET, Rabea P, Ansari SH, Abdin MZ, Sayeed A (2014) Rapid, sensitive, and validated UPLC/Q-TOF-MS method for quantitative determination of vasicine in *Adhatoda vasica* and its in vitro culture. Pharmacogn Mag 10:5198–5205

Mandal J, Laxminarayana U (2014) Indirect shoot organogenesis from leaf explants of *Adhatoda vasica* Nees. Springer Plus 3:1–8

Mathew AS, Patel KN, Shah BK (1998) Investigation on antifeedant and anthelmientic potential of *Adhatoda vasica* Nees. Indian J Nat Prod 14:11–16

Maurya S, Singh D (2010) In vitro callus culture of *Adhatoda vasica*: a medicinal plant. Ann Biol Res 1(4):57–60

Murali RS, Basavaraju R (2012) Influence of plant growth regulators on the in vitro morphogenetic and callogenetic competence of *Adhatoda vasica* Nees. Med Plants-Int J Phytomedicine Rel Inds 4(3):138–42

Murashige T, Skoog F (1962) A revised medium for rapid growth and bioassays with tobacco tissue cultures. Physiol Plant 15:473–497

Nath S, Buragohain AK (2005) Micropropagation of *Adhatoda vasica* Nees–a woody medicinal plant by shoot tip culture. Indian J Biotechnol 4:396–399

Panigrahi J, Gantait S, Patel IC (2017) An efficient in vitro approach for direct regeneration and callogenesis of *Adhatoda vasica* nees, a potential source of quinazoline alkaloids. Natl Acad Sci Lett 40:319–324

Pierik RLM (1987) In vitro culture of higher plants. Kluwer Academic Publishers Group, Dordrecht

Rashmi PA, John Reshma, Mathew Linu (2012) Isolation and characterization of vasicine from in vitro cultures of Justicia adhatoda. Int J Pharm Bio Sci 3:58–64

Roja G, Vikrant BH, Sandur SK, Sharma A, Pushpa KK (2011) Accumulation of vasicine and vasicinone in tissue cultures of *Adhatoda vasica* and evaluation of the free radical-scavenging activities of the various crude extracts. Food Chem 126(3):1033–1038

Sharma N, Gowthami R, Pandey R (2019) Synthetic seeds: a valuable adjunct for conservation of medicinal plants. Synthetic seeds (Germplasm regeneration, preservation and prospect). Springer, pp 181–216

Sil SK, Ghosh MK (2010) Micropropagation through in vitro morphogenesis and embryogenesis of *Adhatoda vasica* Nees. Adv Plant Sci 23(1):11–14

Singh B, Sharma RA (2014) Pyrroloquinazoline alkaloids from tissue cultures of *Adhatoda vasica* and their antioxidative activity. Am J Phytomed Clin Ther 2(3):403–412

Singh B, Sahu PM, Sharma RA (2017) Effect of elicitors on the production of pyrroloquinazoline alkaloids by stimulating anthranilate synthase activity in *Adhatoda vasica* Nees cell cultures. Planta 246(6):1125–1137

Tejavathi DH, Manjula BL, Anitha P (2008) Multiple shoot regeneration from the cultures of *Adhatoda vasica* Nees. In: IV international symposium on acclimatization and establishment of micropropagated plants, vol 865, pp 367–370

Yash M, Amla B, Sharma MM (2015) Histological localization of fungal endophytes in healthy tissues of *Adhatoda vasica* Nees. Curr Sci 112(10):2012–2015

Chapter 6
Phyto-pharmaceutical Potential and the Isolation of Novel Compounds from *Adhatoda vasica* L. Nees

A. vasica contains several bioactive compounds as the derivatives of quinazoline ring in the different parts of the plant (root, stem, leaves, flower, bark, etc.), for instance, vasicine, 5-hydroxy vasicine, vasicine glycoside, deoxyvasicine, vasicinol, vasicinone, vasicolinone, adhavasicinone, adhatodine, vasnetine and anisotine (Dhankhar et al. 2011).

The presence of several phytoconstituents as the metabolites makes the plant effective to be used as a therapeutic agent to treat the diseases. Hence the development of safe and effective medicines are required to go through several steps in suitable orders which initiates at the survey and proper selection of the species, their cultivation and geographical conditions and effects of biotic and abiotic factors (Verma and Shukla 2015; Zaynab et al. 2018) and ends at the phytochemical screenings, quantification of active chemical components which is variable (Parveen et al. 2015) and finally the in vitro and in vivo clinical trials (Liu et al. 2018). Fractional extracts possess the a varied range of constituents, while aqueous extract of the leaves may contain several metabolites as alkaloids, phenols, saponins and tannins (Maurya and singh 2010; Karthikeyan et al. 2009; Chattopadhyay et al. 2011; Jha et al. 2012; Kanthale and Panchal 2014) along with sugars, proteins and glycine as the amino acids. Further relevant studies showed the chief constituent vasicine showing optical activity in normal conditions which gets racemized during extraction similarly as atropine from the leaves of *Atropa belladonna* (Sajeeb et al. 2016).

M. Ali, K. R. Hakeem, *Scientific Explorations of Adhatoda vasica*, SpringerBriefs in Plant Science, https://doi.org/10.1007/978-3-030-56715-6_6

6.1   Crude Extracts of *Adhatoda vasica* and Isolation Protocols

6.1.1   Rationalized Approaches and Suggestions for Phytochemical Analysis

The extraction of phytoconstituents rich crude extract is a tedious job which required suave skills. *Adhatoda vasica* plant is having an enormous potential to have almost all diverse forms of phytochemicals which requires different protocols to follow to achieve the goal. Many research scholar attempted 'trial and error' approach, while others made some changes in the existing ones. Since we know the plant is ancient and had been in use for thousands of years, it reveals the presence of nitrogen containing compounds in good amounts as their chemical identity fingerprints and therapeutic concerns. The different parts of the plant possessing the phyto-chemicals are also responsible to give the considerable amount of chemicals if were used as explants in tissue culture techniques or biotransformation protocols. The various methods which are covered up here also comprise the approaches which were adopted when the sample was a biomass in the form of callus grown under controlled conditions. The different methods which rely totally on extraction and analysis are as follows.

- The leaves (500 g) are required to be shade dried for approx. 2 weeks at the room temperature. Make the fine powder with high-quality stainless steel blender. Now the powder must be processed to be soaked with high purity 99% ethanol (approx. 500 mL) with constant stirring for 24 h. Finally, the extract would be filtered and the filtrate would be concentrated through rotary evaporator at reduced pressure adjusting at 56 °C. The probable yield of dried crude extract must be around 80 g (20% w/w) (Gopalan et al. 2016). The obtained crude extract (approx. 70 g) was dissolved in 700 ml of glacial acetic acid (6%) then warmed at 45 °C. Filter the solution and cool and then basified with liquid ammonia to adjust pH 8–9. Extract with chloroform 3 times and wash the resulting chloroform layer with water and saturated solution of sodium chloride. Processed it to dry over sodium sulphate, concentrate under vacuum dryer under reduced pressure to get the optimum yield approx. 12 g which was calculated as 12% w/w (Joshi et al. 1994). This yield was further projected for column chromatography to have the pure isolates of components. The mobile phase needs to be standardized, which might be varied in terms of composition as per the provided physical conditions.
- Dried plant powder of *A. vasica* (1.5 kg) was cold macerated exhaustively and extracted sequentially with increasing polarities of the organic solvents, i.e. 3 L of hexane and ethyl acetate and then methanol (at normal room temperature). Intermittent shaking was carried out for 48 h in soaking state. The resultant extracts were obtained, collected and concentrated by using rotary evaporator developing the reduced pressure maintaining approx. 40 °C temperature. The

same was further concentrated in vacuum rotary evaporator at the same temperature and finally stored at 4 °C for further use. This extract was subjected for the pure isolation through column chromatography (Ignacimuthu and Shanmugam 2010).

- Leaves were kept for shade drying for 21 days and they made a fine powder with grinder. Double distilled water (deionized) was use to soak the fine powder in the ratio of 1 g:100 ml. It was then processed for sonication for half an hour then centrifuged. Supernatant solution was separated and the next day was again centrifuged. Resultant solution was purified using Whatman filter paper no. 1 (Bhavyasree and Xavier 2020).

- This extract was prepared for antioxidant studies and HPLC estimations where 1 g of shade dried powdered was mixed with 10 ml of distilled water. It was thoroughly mixed with the help of a mechanical shaker and then filtered through Whatman filter paper. The clear solution was stored in the refrigerator for further use (Ahmed et al. 2018).

- Cold percolation method was processed to the shade dried coarsely powdered leaves of *A. vasica* in ethanol. Percolation was carried out for 48 h by maintaining the solvent volume. Shade dried amount of leaves 2 kg was calculated finally to yield 100 g of the final extract after filtration through Whatman filter paper and distillation.

- Column chromatography was done by using the small proportion of the above yield. The column was run by using chloroform and the mixture of chloroform and methanol in the ration of 4:1. The elution result confirmed the presence of vasicine as the single spot by using the chloroform and methanol mixture. Its purity was examined by its melting point and spraying it with Dragendorff's reagent having orange colour.

- Shade dried plant material was pulverized and dipped in ethanol to have the vasaka extract. The extract was obtained by filtration and then dissolved in 5% acetic acid and kept at 60 °C for 15 min. Defatting was done by using nonpolar organic solvent, i.e. n-hexane. The leftover remaining aqueous layer was basified with ammonium hydroxide, and then extraction was carried out with chloroform to have vasicine which was purified and recrystallized with methanol-acetone of equal compositions (Shahwar et al. 2012).

- An adequate amount of shade dried and finely powdered *Adhatoda* leaves were kept overnight in ethanol (95%). Fresh ethanol of the same strength was again used to re-suspend the filtered residue for 48 h then filtered. The whole ethanolic filtrate was mixed and evaporated to half at reduced pressure by using rotary evaporator. The semisolid yield was kept at – 4 °C for further experimental uses (Manoharan and Prabhakar 2014).

- Capillary electrophoresis as a novel approach was employed to determine vasicine and vasicinone quantitatively from *A. vasica* plant. Fused silica capillary was used to separate the content in approx. 11 min with good reproducibility parameters and finally was validated (Avula et al. 2008).

- Air-dried powdered leaves were extracted with methanol by using Soxhlet apparatus. The methanolic extract was concentrated under vacuum reduced pressure to get a changed colour gummy residue. The methanolic extract was then acidified with 5% hydrochloric acid and extracted with organic solvent chloroform to remove the impurities and non-alkaloidal content. The acidified fraction was maintained by basifying with ammonia, and the pH was adjusted to 9–10 to have an efficient extraction with chloroform by using the separating funnel in several steps. Repetition of the steps was to extract the whole alkaloidal content from aqueous phase. Aqueous phase was having some other metabolites, i.e. saponins which were separated by following the method published by Hariharan and Rangaswami (1970). All three fractions were obtained by evaporation to dryness. The fractions were marked as the alkaloidal fraction, non-alkaloidal and isolated saponins fraction (Chakraborty and Brantner 2001).

- Extraction of the content for chromatographic and phytochemical analysis required air-dried (30–50 °C) coarsely powdered leaves, stems, and roots of *A. vasica* with almost equal quantities. The extraction was carried out in separate flask with methanol containing 20% v/v of ammonium hydroxide. The extracts were then concentrated under vacuum and re-dissolved in methanol to get filtered and re-concentrated purified extracts which are to be followed for HPTLC analysis (Das et al. 2005).

- The process to extract the vasicine exclusively was mentioned by Chattopadhyay et al. (2003) and patented in 2003 as well, which includes the steps presented in the Fig. 6.1, where shade dried pulverized leaves of *A. vasica* were processed to be extracted out with alcohol by using the Soxhlet apparatus. Aqueous organic acid (citric acid) was mixed and stirred for 2–24 h. Acid solution was extracted by using preferably dichloromethane in separating funnel. Aqueous acidic solution was basified with liquid ammonia, and the basified solution was now separated by using organic solvent chloroform. The organic layer was separated, dried and filtered. Chloroform was evaporated to get the amorphous residue which was then dissolved with the mixture of petroleum ether and acetone in binary or 2:1 compositions, respectively. The purity was revealed later on as 80% which needs to be checked again to improve in terms of purity and yield (Chattopadhyay et al. 2003). The yield of vasicine also depends upon the time of collection of the leaves due to environmental conditions. Pandita et al. (1983) reported the best availability of the vasicine in August to October in the leaves of *A. vasica*. The total alkaloidal content was believed to be comprised about 95% of vasicine. Seasonal variation was also reported by Bagchi et al. in 2003 who claimed it to be high during the month of March collection.

- Vasicine which was isolated earlier from *A. vasica* leaves can easily be used to synthesize the derivative as vasicine acetate to meet the enhanced antimicrobial properties. Vasicine acetate was found to show the moderate antibacterial activity in comparison to pure vasicine. The required steps are well illustrated in Fig. 6.2 (Duraipandiyan et al. 2015).

- Minor modifications in the steps were always been fascinating attempts to improve the quality of the content. The procedure which was mentioned by Atal (1980) was to suggest the use of 95% alcohol at the very first step of extraction through Soxhlet by using the shade dried leaves of *A. vasica*. The alcoholic

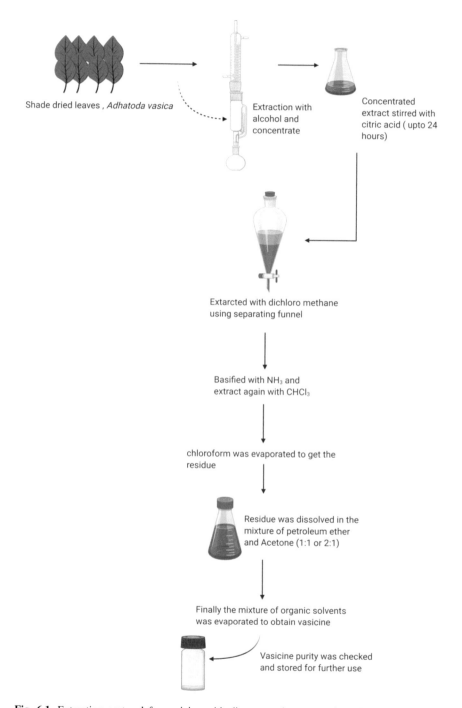

Shade dried leaves , *Adhatoda vasica*

Extraction with alcohol and concentrate

Concentrated extract stirred with citric acid (upto 24 hours)

Extarcted with dichloro methane using separating funnel

Basified with NH$_3$ and extract again with CHCl$_3$

chloroform was evaporated to get the residue

Residue was dissolved in the mixture of petroleum ether and Acetone (1:1 or 2:1)

Finally the mixture of organic solvents was evaporated to obtain vasicine

Vasicine purity was checked and stored for further use

Fig. 6.1 Extraction protocol for vasicine with diagrammatic presentation. (Image is prepared through biorender.com)

Vasicine (1g) + acetic anhydride (25 mL)
The flask was kept for overnight in the fridge

↓

Next day 5 drops of conc. Sulphuric acid were added to
the above flask, stirred and kept n water bath at 50°C for half an hour

↓

Kept over broken ice bed and diluted with 250 mL of water

↓

Extracted with chloroform twice with 250 mL, in a separating funnel
And extracts were combined

↓

Washed with water and dried over anhydrous sodium sulphate and distilled
Finally the residue was washed with less amount of n-hexane

↓

Residue was crystallized from chloroform to have vasicine acetate

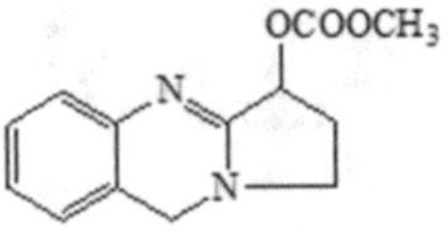

OCOOCH₃

Vasicine acetate

Specifications-
Colorless crystal of vasicine
acetate, m.p. 120.

TLC single spot at R_f =0.6
Solvent system – n-hexane: ethyl
acetate (2:1), upon spraying
Dragendorff's reagent – **orange
red color**

Fig. 6.2 Acetylation of vasicine to synthesize vasicine acetate for enhanced antibacterial
properties

extract was concentrated and treated with aq. sulphuric acid (2%) which was later on basified with liquid ammonia to be extracted with organic solvent chloroform. The chloroform layer was separated by using the separating funnel and was dissolved in aqueous sulphuric acid with 2% strength. The process of basification was repeated with liquid ammonia to be extracted again with chloroform. The procedure with few shortcomings was also monitored critically and found the use of strong mineral acid (sulphuric acid) not suitable for the vasicine to maintain its properties. There were chances of considerable degradation of the compound (Atal 1980).

- *A. vasica* plant gained the global attention in date back to history. Mehta et al. (1963) have also been reported a procedure mentioning the complete protocols. The Soxhlet was not used for the alcoholic extraction of the leaves, but this time it was directly refluxed with 90% alcohol. The solvent was finally evaporated to have the concentrate of residues which was later on dissolved in hot distilled water. The extract was filtered out and again extracted with chloroform to remove any impurity or if left colouring matters. The reaming aqueous layer was basified with sodium hydroxide (caustic soda) 5% and again was extracted with chloroform. This chloroform extract was acidified with hydrochloric acid (5%) and sequentially basified again by using different base as liquid ammonia. Finally it was processed to be extracted with chloroform several times and was concentrated to give total alkaloid where vasicine was obtained as vasicine hydrochloride with considerable yield. The hot water addition was revealed to have two drawbacks as difficulties in quantitative extraction of vasicine and the conversion of vasicine into vasicinone through auto oxidation process (Kumar et al. 2014).

- When the quinazoline alkaloids (vasicine, vasicinone and derivatives) are to be extracted from other plant sources, the sources are required to be dried and cut into pieces to be extracted with 50% ethanol (v/v) by refluxing for 2 h. The repeated process provides a good amount of content to be filtered and finally concentrated under reduced pressure about 45 °C. The solution of this concentrated extract would be used to prepare the different alkaloid fractions by utilizing macro porous resin column chromatography (Liu et al. 2015).

- Fresh leaves are required to be washed with distilled water (d.w.) just prior to be shade dried at least for 2 weeks of durations at room temperature. The shade dried leaves needed to be ground into coarse powder. The aqueous extract is to be prepared by adding mentioned quantity of leaf powder of *A. vasica* in distilled water. The mixture is required to be heated below the boiling point and filtered through the Whatman no.1 filter paper. This leaf extract is to be processed for the synthesis of gold nanoparticles (Latha et al. 2018).

- Decoction method was also used to prepare the aqueous extract of *A. vasica*. The process includes the fresh collection of the leaves and its drying under shade to avoid any deleterious effect on thermo-labile and essential components. The washing and drying is subjected to be cut into pieces and mix with hot water to be boiled at this stage for an hour. Cooled mixture was filtered using a Whatman filter paper, and the obtained filtrate was evaporated to have a concentrated decoction. It was then lyophilized to powder and stored in amber coloured boro-silicate airtight vials (Wilson et al. 2020).

- Extraction by using root cuttings is required to defat with hexane to finely extract with ethanol. The alcoholic extract then needed to be acidified with 5% hydrochloric acid (HCl) and extracted with chloroform to remove impurities and other non-alkaloidal components. The acidic extract would be required to basify with solution of ammonia by adjusting the pH (9—10), and finally it would be extracted with chloroform to provide the crude alkaloids, which later on would be purified over silica gel (Jain and Sharma 1982).

- As per the method for extraction of phyto-components, the shade dried and coarsely powdered leaves of *Adhatoda vasica* were extracted with ethanol. The repetition was done for the cold percolation procedure for about 48 h. Finely the extract was filtered by using Whatman no.1 filter paper and later on distilled on a water bath. It was then concentrated to a dark coloured green residue which subsequently was dried in vacuum (Duraipandiyan et al. 2015).

- Aqueous extract of *A. vasica* was prepared (to detoxify the aflatoxins) by homogenizing 1 g of plant material (leaves/seed) in 3 mL of sterile distilled water, subsequently to be centrifuged at $14000 \times g$ for 15 min. Finally supernatant was used for the study purpose (Vijayanandraj et al. 2014).

- Aqueous suspension of dried leaves was prepared as paste to extract the sugars (polysaccharides) by stirring it water with maintained pH 6 at 25–32 °C for 12 h, later on filtered through glass filter (G-2) and residue was repeated again to have the dissolved components in the ratio of 1:100 (w/v), subsequently were dialyzed against water and lyophilized. Ethanol was used to precipitate, and repeated centrifugation gave the final pellets which were re-dissolved and lyophilized to have the water extracted polysaccharides (Chattopadhyay et al. 2011).

- Air-dried fresh leaves of *A. vasica* were grounded and dipped in methanol at room temperature for 72 h. The repetition was done three times and combined extract collected then to be filtered and concentrated under reduced pressure to yield 4.7% of fresh weight. One part of the combined total extract was suspended in distilled water and then acidified to the adjusted pH of 2–3 using 2% sulphuric acid to be partitioned with dichloromethane (DCM). The acidic solution was then neutralized with liquid ammonia (pH 10–12) and again extracted with DCM. The dichloromethane was evaporated to have a significant yield of semi-purified alkaloidal fraction (Ali et al. 2016).

- Vasicine and vasicinol as the active principles were isolated from *A. vasica* leaves extract to evaluate the inhibitory effect on α-glucosidase enzyme to support the anti-diabetic potentials. Aqueous methanolic extracts were fractionated and subjected to the Diaion HP-20 CC (column chromatography) to have many fractions. Active fraction was identified on the basis of inhibitory effectiveness and separated by Cosmosil $75C_{18}$-OPN column chromatography (CC). Final fraction was further purified by Sephadex LH-20 CC column chromatography to yield alkaloidal components vasicine and vasicinol (Gao et al. 2008).

- Soxhlet fractional method was reported by Yadav et al. (1992) to extract the active crude extract to further evaluate the anticestodal activity of *A. vasica*.

Shade dried (to protect the heat labile components) leaves were powdered, and methanol was used to extract at 40 °C through Soxhlet apparatus. Final crude extract was processed to evaporate to dryness and store at 8 °C for further use. Phytochemical-rich extract yielded 1.67% (w/w) of the shade dried weight (Yadav and Tangpu 2008).

- Alcoholic and hydroalcoholic extracts were prepared by using Soxhlet apparatus under controlled adjustment of the conditions, filtered through Buchner with Whatman filter number no.1 and concentrated under reduced pressure using rotary evaporator at 50–55 °C, dried in air or incubator (Saha and Bandyopadhyay 2017). Stock solution of 1% strength was prepared in the respective solvents for further use (Saha and Bandyopadhyay 2020).

- Shade dried powdered roots of *A. vasica* were soaked for maceration in ethanol for 2 weeks and filtered. Filtrate was concentrated to gum at reduced pressure; then the gummy extract was treated and acidified with 5% hydrochloric acid, warmed and filtered again to be defatted with hexane. After defatting the extract was basified with ammonia to adjust the pH at 10. Basified extract was fractionated with chloroform, and now it was having the mixture of quinazoline alkaloids which was purified and characterized by column chromatography (Joshi et al. 1994).

- Ultra-pure water was used to wash the fresh leaves to be processed for shade drying for the purpose of Soxhlet extraction in hydro-alcoholic solvent. Temperature was adjusted to 60 °C. The extract was cooled and filtered through Whatman filter paper no. 1, concentrated and kept at 4 °C for further use (Pandiyan et al. 2018).

- Shade dried powdered drug *A. vasica* was soaked in methanol in conical flask and incubated at 37 °C with regulated 120 rpm per 30 min for a day and night. Filtered by using gauze having four layers and then with Whatman filter no. 1. The same steps were repeated three times to have complete phytochemical extraction. The collected filtrates were pooled and evaporated to dryness using desiccators and finally reconstituted in absolute ethanol to be estimated for total concentration of quinazoline alkaloids (Pa and Mathew 2012).

- Areal parts (leaves and stems) of *A. vasica* were dried in oven at 60–70 °C for 24 h and was powdered in mixer to be defatted with petroleum ether and filtered out. Petroleum ether extract was evaporated and analysed. The left residue was dissolved and extracted with methanol to analyse the alkaloids. The methanolic extract was dissolved in water and basified with strong alkali (NaOH) with adjusted pH 12 and then extracted three times with chloroform; the basic fraction was obtained to be processed for chromatographic analysis. The yield of marker compounds vasicine and vasicinone in various extracts of areal parts (leaves and stem), and their tissue culture bio mass (Roja et al. 2011) has been summarized in Table 6.1.

- Callus obtained from leaves and stems was dried and suspended in methanol, stirred and filtered. The filtrate was concentrated and partitioned between ethyl acetate and HCl (1N). Sodium bicarbonate was used to neutralize, and finally ethyl acetate was further added as suggested by Saxena et al. (1997). Ethyl

Table 6.1 Quantified percentage of quinazoline marker compounds in areal parts and callus grown by using them

S. No.	Marker compounds (Percentage dry weight)		Extracts
	Vasicine	Vasicinone	
I.	0.187	Traces	Areal parts, petroleum ether extract
II.	0.037	2.19	Tissue culture, petroleum ether extract
III.	2.80	Traces	Areal parts, methanolic extract
IV	1.40	1.96	Tissue culture, methanolic extract
V	3.64	3.40	Areal parts, water extract
VI	5.98	5.20	Tissue culture, water extract

acetate fraction was evaporated and titrated again with ethyl acetate or dichloroethane. The presence of alkaloids was revealed by using phytochemical evaluation (Jayapaul et al. 2005).

- Shaded dried crushed powdered leaves of *A. vasica* were allowed to be soaked in 95% ethanol in separating funnel for a complete day night with occasional shakings. Then it was filtered through Whatman no. 1 filter and concentrated to dryness using rotary evaporator under reduced pressure. The final extracted dried content was freeze-dried and stored at 4 °C for further use (Patil et al. 2014).

- Areal parts (leaves and flowers) were washed and dried at room temperature under shaded area. The dried sample was coarsely powdered and macerated with 80% ethanol for 3 h and filtered by using Whatman no. 1 filter paper. The filtrate was then kept for drying at 40 °C for 1 week in dark cool and dry place (Dhuley 1999).

- Callus grown dried powdered cells were sonicated with methanol and were centrifuged. Supernatant was taken out and repetition was done to combined all supernatants. Final volume was adjusted as per reference in methanol and processed for HPLC analysis (Bhambhani et al. 2012).

- Simple cold extraction was done by putting shade dried leaves in methanol with constant stirring for 24 h. The filtrate was concentrated under reduced pressure at 50 °C by using vacuum dryer. The dried extract was finally mixed with 10% dimethyl sulfoxide and kept at 4 °C for further use. Methanol was also used as a solvent for another method of extraction by Soxhlet apparatus. It was run for 8 h for complete extraction. The filtrate was concentrated under reduced pressure at 40 °C by using vacuum dryer. The dried extract was finally mixed with 10% dimethyl sulfoxide and kept at 4 °C for further use (Jha et al. 2014).

- Leaves of *A. vasica* were made into pieces to be extracted with 5% acetic acid through maceration process. It was then filtered and filtrate was dried at 70 °C and concentrated. The basic pH at 10 was adjusted with ammonium hydroxide and was centrifuged. The precipitate was washed again with ammonium hydroxide and centrifuged. The crude extract was dissolved in hot boiling methanol and processed to be recrystallizing the alkaloids. Finally it was stored at 4 °C for further use (Jha et al. 2012).

- 2 g of shade dried leaf powder was allowed to mix with 50 ml of ethanol (95%) at 45 °C in ultrasonic device of extraction for half an hour. The process was repeated for three times to have the better yield. The extract was finally collected, filtered and dried at 50 °C using the rotary evaporator. Moreover other methods also follow the different approach as methanol (70%) was used to soak the finely ground dried leaves for 12 h with shaking. Methanol was evaporated under sterile conditions to have the crude extract of phytochemicals. Centrifugation was also done in some different ways to have the supernatant and then to decolorize it with addition of charcoal and finely air-dried (Adhikary et al. 2016).

The procedures and adopted protocols for the extraction of crude form of the phytochemical residues and isolations of pure compounds have been mentioned for the sake of further explorations to make it easy and reproducible outcomes. The procedures and the chromatographic analysis along with their developed solvent system have been mentioned in Table 6.2, which helps to suggest and to validate the techniques with minor efforts by saving the time.

6.2 Phytochemical Constituents (Primary and Secondary Metabolites)

Quinazoline alkaloids, i.e. vasicine and vasicinone, are known to have quinazoline ring. The derivatives of the compounds are derived from this ring containing compounds, and the ring is biosynthetically synthesized by taking up the primary metabolites such as various amino acids (glutamic acid, proline, ornithine, etc.) following the shikimic acid pathway contribution as well. The biosynthetic steps are figured out in Fig. 6.3.

Vasicine (1,2,3,9-tetrahydropyrrolo[2,1-b]quinazolin-3-ol, $C_{11}H_{12}N_2O$; Mol. Wt.: 188.226), off white powder, $[\alpha]_D^{25} -125^0$(Gao et al. 2008), alkaloid is a heterocyclic biomarker compound present in the leaves of *A. vasica* having quinazoline nucleus which is primarily present in the leaves of *Adhatoda vasica*, family Acanthaceae, and in roots of two *Sida* species (*Sida cordifolia* and *Sida acuta*) and the different parts of *Peganum harmala* (Zygophyllaceae) (Dhalwal et al. 2010; Rajani and Kanaki 2008). Vasicine as being the light-sensitive compound gets easily auto-oxidized to vasicinone upon exposure to light and air (Brain and Thapa 1983; Chowdhuri and Hirani 1987). Synthetically the marker compounds are also prepared by adopting different chemical schemes. Schopf and Oechler (1936) scheme is very famous to yield vasicine from 2-aminobenzaldehyde and γ-amino-α-hydroxybutyraldehyde (Leonard and Martell 1960; Fig. 6.4). There were several modifications in the schemes to have better yield of the alkaloidal components (Wasserman and Kuo 1991). Vasicine being quinazoline alkaloid has also been utilized a proven and efficient catalyst to have metal and base-free Henry reaction of many aldehydes along with nitro alkanes. Vasicine is also utilized to synthesise vasicinone isomers (l&d) by using Davis reagent [2-(Phenyl sulphonyl) -3- phenyl –

Table 6.2 Solvents system developed for the chromatographic analysis

S. No.	Part used	Solvent system	Chromatographic analysis	Reference
(i)	Leaves, stem and calli	Toluene/butanol/butyl acetate (9:0.5:0.5; v/v/v)	High-performance thin layer chromatography (HPTLC)	Panigrahi et al. (2017)
(ii)	Leaves powder (Hexane extract)	Hexane/ethyl acetate (8:2)	Column chromatography (CC)	Ignacimuthu and Shanmugam (2010)
(iii)	Methanolic extract	Acetonitrile (0.1 M)/ phosphate buffer/glacial acetic acid (15:85:1) (pH adjusted 4.0 by phosphoric acid)	High-performance liquid chromatography (HPLC)	Subramanya et al. (2016)
(iv)	Fraction extracted in alcohol (50%)	Methanol– dichloromethane–ammonia mixtures (3:2:0.1)	Column chromatography (CC)	Abdullaev et al. (1983)
	For quinazoline alkaloids			
(v)	Extracts of leaves, stem and roots of A. vasica	Methanol/toluene/dioxane/ ammonia (2:2:5:1	High-performance thin layer chromatography (HPTLC)	Das et al. (2005)
(vi)	Alcoholic extract of leaves	Pure chloroform	Column chromatography (CC)	Duraipandiyan et al. (2015)
		Chloroform and Methanol mixture (4:1)		
(vii)	Alcoholic extract of leaves	Chloroform/methanol (3:1)	Thin layer chromatography (TLC)	Duraipandiyan et al. (2015)
		For single spot vasicine		
(viii)	Areal parts (leaves and stem) methanolic extract	Dichloromethane/methanol/ water (70:25:5)	Thin layer chromatography (TLC)	Roja et al. (2011)
(ix)	Areal parts (leaves and stem) methanolic extract	Methanol/water (40:60)	High-performance liquid chromatography (HPLC)	Roja et al. (2011).
(x)	Ethyla cetate extract of Callus	Methanol/chloroform (1:9)	Thin layer chromatography (TLC)	Pandita et al. (1983)
(xi)	Dried and powdered cells (callus from areal parts)	Methanol/water (40:60)	High-performance liquid chromatography (HPLC)	Bhambhani et al. (2012)

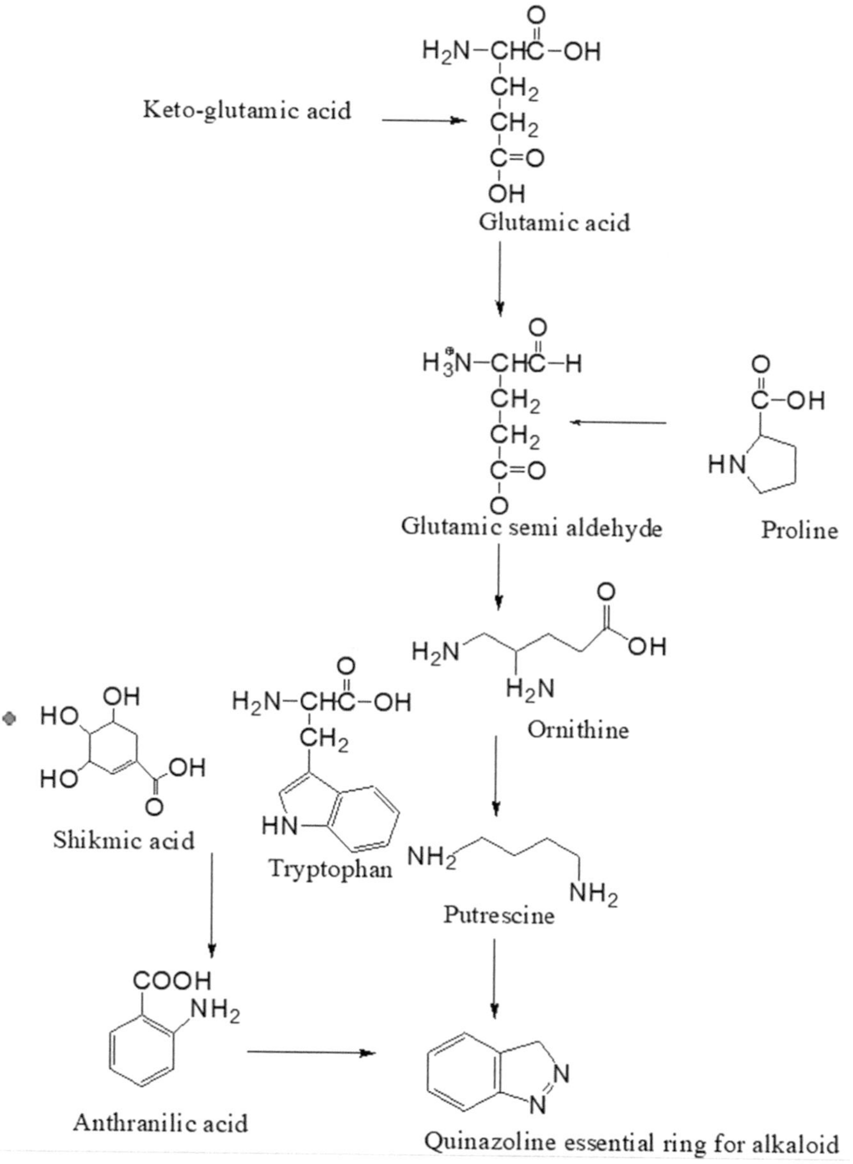

Fig. 6.3 Biosynthesis of quinazoline ring by using amino acids and shikimic acid

oxaziridine], Fig. 6.5. The utility favours to be used for the synthesis of several β-nitro alcohols under mild conditions of reaction without using the toxic and hazardous organic solvents and many expensive catalysts as well (Sharma et al. 2016).

Vasicine is also known as peganine (the important phytoconstituent of *Peganum harmala*) which is important alkaloid. Other constituents in the considerable amounts are adhatodic acid and other quinazoline alkaloids. The vasicine

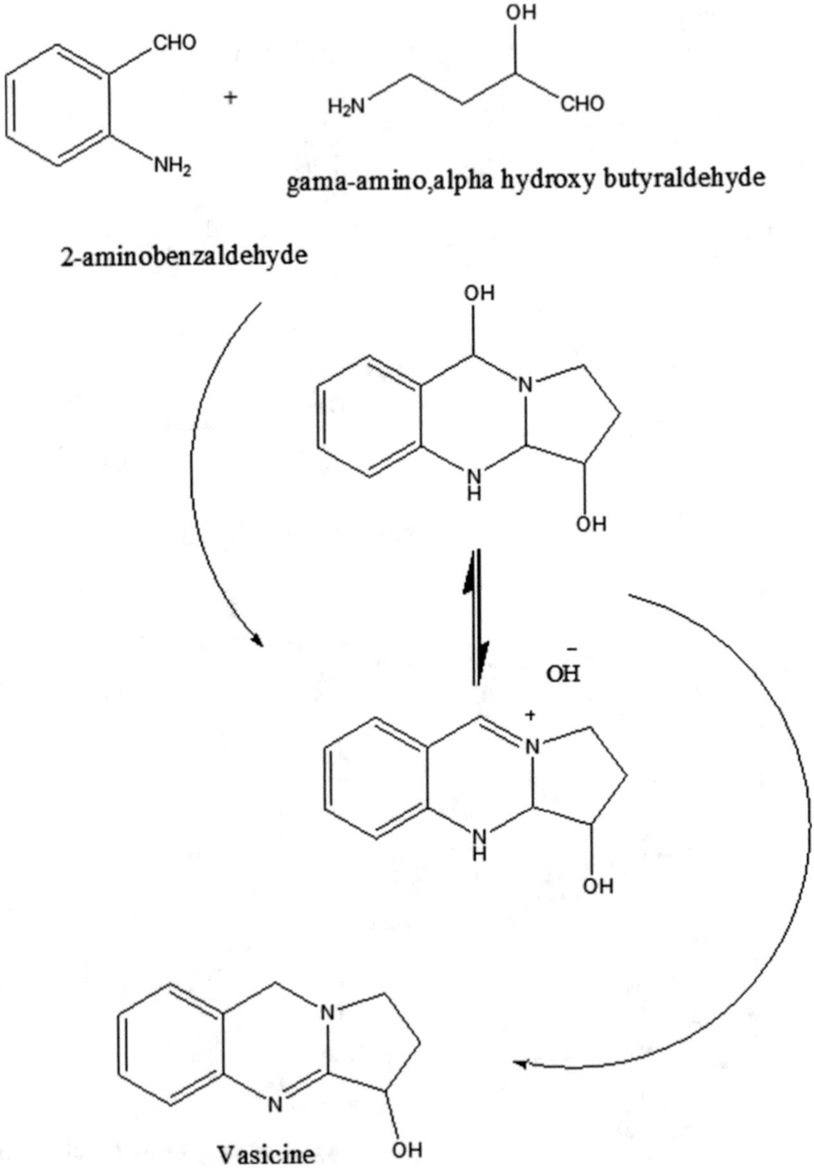

Fig. 6.4 Synthesis of vasicine through Schopf and Oechler scheme

(peganine) constitute the major proportion as 0.25% within alkaloids. Major components as quinazoline ring containing alkaloids, i.e. vasicine, vasicinone, vasicoline, vasicolinone, adhatodine, anisotine, hydroxyl peganine and steroids, have already been isolated from the leaves (Jha et al. 2012; Lahiri and Prahdan 1964). Other reported constituents are vasakin, anisotinine, vasicinine, betaine and vasicinol in the plant (Ikram and Ehsanul Haq 1966; Johne et al. 1971). Dymock

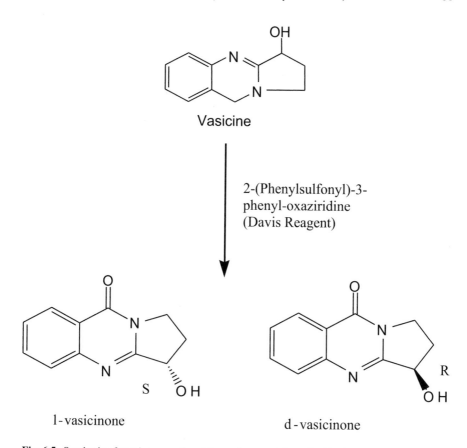

Fig. 6.5 Synthesis of two isomers of vasicinone from vasicine using Davis reagent

et al. (1893) reported their findings in leaves as it contains volatile odorous principle 0.2%, alkaloid 3.2% with fat and miscellaneous others, sugars and resin 12.5%, colouring matter 4.83%, gum 3.87%, other organic substances and salts 10.38%, inorganic residue 9.59% and organic residue 40.71%. Ash values contain water-soluble components 23.38%, acid-soluble components 75.12% and insoluble substances as 1.5% (Akbar 2020). The presence of volatile oils, resins, fats, sugars, gum, proteins, amino acids and vitamins 'C' has also been reported (Bhat et al. 1978). The phytochemical analysis further showed the presence of alkaloids, phenolic components and flavonoids with less amount of tannins, saponins, anthraquinones and some reducing sugars as well (Pathak 1970).

The mineral elements were screened to reveal the presence of macro and micro-elements. Potassium, calcium, iron, copper, zinc, chromium, vanadium and manganese were detected in the extract of leaf sample. Calcium showed the highest concentration as (68070± 35.58 ppm) (Kumar et al. 2014). Other reporting also revealed the leaves to contain pyrroloquinazoline alkaloids, chiefly vasicine (1,2,3,9– tetrahydropyrrolo[2,1–b]quinazolin–3–ol, $C_{11}H_{12}N_2O$), other alkaloids as

9-acetamido-3,4-dihydropyrido-(3,4-b)-indole,   oethyl-A-D-galactoside-1,2,3,9-
tetrahydropyrrolo(2,1-b) quinazolin-9 (1H)-one and deoxyvasicinone, vasicol
(1,2,3,4,9, 11-hexahydropyrrolo(2,1-b) quinazolin-3,11-diol. Vasicinolone was also
reported as the oxidative product of vasicinol (Khursheed et al. 2010). The phytochemical screening was also carried out for all the different parts and reported by
Khursheed et al. (2010). Table 6.3 summarized the respective constituents with
corresponding parts of the plant. The floral parts were constituting of flavonoids

Table 6.3 Phytochemical screening through different parts of the *A. vasica* L.Nees

S. No.	Extracts and chemical categories	Chemical constituents
(i)	Alkaloids – ethanolic extract (leaves)	Desmethoxy aniflorine
		7-Methoxy vasicinone
(ii)	Petroleum ether extract (leaves)	29-Methyl triacontan-1-ol (aliphatic alcohol)
	Non-nitrogenous compounds	
(iii)	Petroleum ether extract (flower), non-nitrogenous compounds	Triatricontan
		β-Sitosterol
		β-Sitosterol-D-glucoside
		α-Amyrin
(iv)	Ether extract (flower)	Kaempferol
		Quercetin
(v)	Ethyl acetate and n-butanol extract (Flower)	Kaempferol-3-β-D-glucoside
		Kaempferol-3-sophoroside
(vi)	Fat containing traces (flower)	Tridecanoic acid
		Pentadecanoic acid
		Novel glucoside – 2′4-dihydroxy-chalcone-4-glucoside
(vii)	Flavonoids (leaves)	Kaempferol
		Quercetin
		Vitexin
		Isovitexin
(viii)	Phenolic (leaves)	p-Hydroxy benzoic acid
		Syringic acid
		p-Coumaric acid
(ix)	Vitamins (leaves)	Free vitamin C and carotene
(x)	Flavones (leaves and flower)	Luteolin
(xi)		
(xii)	Lignin (Non-woody leaf stalk)	Guaiacyl-, syringyl- p-hydroxy phenyl propane
(xiii)	Fatty acids – fixed oil (seed)	Arachidic acid (3.1%)
		Behenic acid (11.2%)
		Lignoceric acid (10.7%)
		Linoleic acid (12.3%)
		Oleic acid (49.9%)
(xiv)	Unsaponifiable matter fixed oil (seed)	β-Sitosterol

(apigenin, astragalin, kaempferol, quercetin and vitexin), triterpenes (α-amyrin) and alkanes (Haq et al. 1967) and vasicinone (3–hydroxy–2,3-dihydropyrrolo[2,1–b] quinazolin– 9(1H)–one, $C_{11}H_{10}N_2 O_2$) (Amin and Metha 1959). Vasicoline, vasico-linone, vasicinol, adhatodine, adhatonine and anisotine (Johne et al. 1971). The novel alkaloidal components from the leaves were characterized as 1, 2, 3, 9-tetrahydro-5-methoxypyrrol [2, 1-b] quinazolin-3-ol and adhavasinone (Chowdhury and Bhattacharyya 1985). A new moiety as 2′,4-dihydroxy chalcone 4-glucoside was also been identified in floral parts (Bhartiya and Gupta 1982), while vasaka roots were containing deoxyvasicine, sitosterol and β-glucoside-galactose (Iyengar et al. 1994). Water extracted fraction was also been reported to have polysaccharides and other sugar and protein components as uronic acid, pro-tein, rhamnose, arabinose, xylose, mannose, galactose and glucose (Chattopadhyay et al. 2011). Aliphatic hydroxyl ketones 37-hydroxyhexatetracont-1-en-15-one and 37-hydroxyhentetracontan-19-one were also detected on mass spectral basis (Singh et al. 1991).

Various synthetic analogues of vasicine were also synthesized and characterized as 6, 7, 8, 9, 10, 12-hexahydro-azepino-[2, 1-b] quinazolin-12-one, 7,8,9,10- tetra-hydroazepino [2,1-b]quinazolin-12(6H)-one, 6,7,8,9,10-hexahydroazepino[2,1-b] quinazoline-12(5H)-one, 2,4-dibromo-7,8,9,10-tetrahydroazepino[2,1-b] quinazolin-12(6H)-one, 2,4-dimethoxy-7,8,9, 10-tetrahydroazepino[2,1-b] quinazolin-12(6H)-one,2,4-dihydroxy-7,8,9,10-tetrahydroazepino[2,1-b] quinazolin-12(6H)-one, 5-methyl-5a,6,7,8,9,10- hexahydroaze-pino[2,1-b] quinazoline-12(5H)-one, 5-ethyl-5a,6,7,8,9,10-hexahydroazepino[2,1-b]- quinazoline-12(5H)-one,5-benzoyl-5a,6,7,8,9,10-hexahydroazepino[2,1-b] quinazoline–12(5H)-one, 5-benzyl-5a, 6, 7, 8, 9,10-hexahydroazepino [2,1-b] quinazoline-12(5H)-one,5-acetyl-5a, 6, 7, 8, 9, 10-hexahydro azepino [2,1-b] quinazo-line-12(5H)-one(Rayees et al. 2014). Secondary metabolites (quinazoline and steroidal ring containing compounds daucosterol, epitaraxerol) and their syn-thetic and semisynthetic derivatives are also illustrated in Fig. 6.6.(Singh et al. 2011).

6.3 Structure Activity Relationship (SAR) of Vasicine

Ring A substitution in Fig. 6.7 doesn't show any positive change towards the phar-macological response of the compound (Atal et al. 1979). Modification in ring B and the removal of hydroxyl (-OH) group from next ring show a better and highly potent compound deoxyvasicinone which is better bronchodilator than normal vasi-cine (secretolytic analogue of ambroxol (Gibbs 2009). Modification in the ring C makes the compound with altered response of activities. A considerable change in bronchodilatory potential was observed when the ring size was extended up to three additional carbons, i.e. the dihomo analogue. The dihomo analogue was proved to be highly potent bronchodilator of the series. Further increase led to the conclusion of decreased activity. The seven-membered modified analogue was found to be 6–10 times more efficacious than aminophyline subjected to the calculated dose.

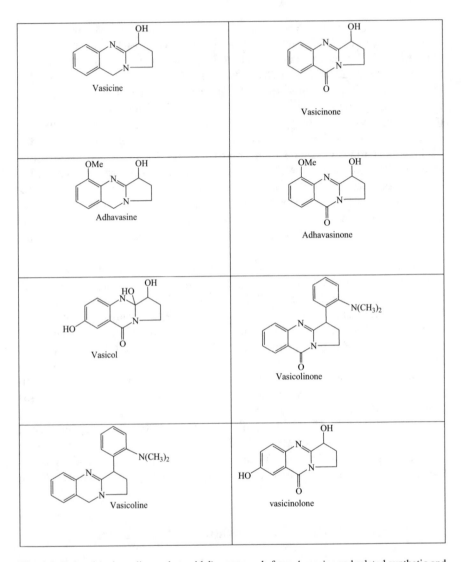

Fig. 6.6 Isolated (quinazoline and steroidal) compounds from *A. vasica* and related synthetic and semisynthetic derivatives

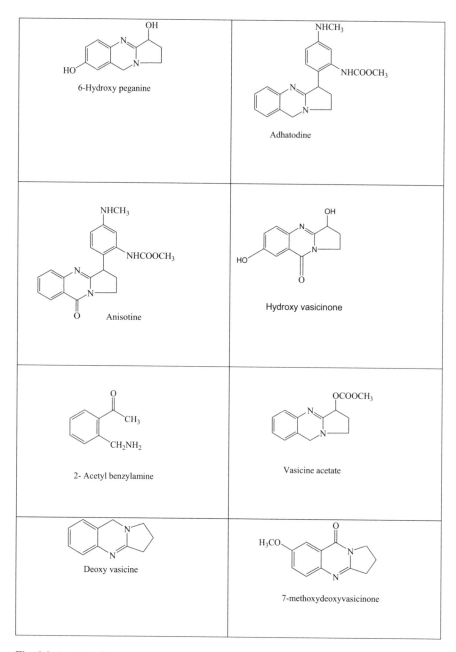

6-Hydroxy peganine

Adhatodine

Anisotine

Hydroxy vasicinone

2- Acetyl benzylamine

Vasicine acetate

Deoxy vasicine

7-methoxydeoxyvasicinone

Fig. 6.6 (continued)

Epitaraxerol

Daucosterol

Bromhexine

Ambroxol

Sputolysin

Peganidine

Fig. .6.6 (continued)

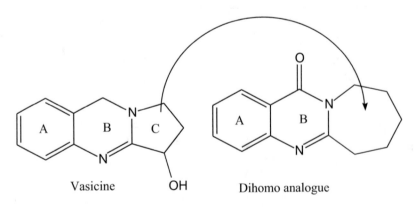

Vasicine

Dihomo analogue

Fig. 6.7 Vasicine to show Structure activity relationship (SAR)

References

Abdullaev ND, Rashkes YV, Yagudaev MR, Tulyaganov N (1983) Structure of the products of the metabolism of deoxypeganine and of deoxyvasicinone. Chem Nat Comp 19(6):720–727

Adhikary R, Majhi A, Mahanti S, Bishayi B (2016) Protective effects of methanolic extract of *Adhatoda vasica* Nees leaf in collagen-induced arthritis by modulation of synovial toll-like receptor-2 expression and release of pro-inflammatory mediators. J Nutr Intermed Metab 3:1–11

Ahmed W, Azmat R, Khan SM, Khan MS et al (2018) Pharmacological studies of *Adhatoda vasica* and *Calotropis procera* as resource of bio-active compounds for various diseases. Pak J Pharm Sci 31(5):1975–1983

Akbar S (2020) Handbook of 200 medicinal plants. Springer, Cham, pp 1059–1066

Ali SK, Hamed AR, Soltan MM, El-Halawany AM, Hegazy UM, Hussein AA (2016) Kinetics and molecular docking of vasicine from *Adhatoda vasica*: an acetylcholinesterase inhibitor for Alzheimer's disease. S Afr J Bot 104:118–124

Amin AH, Metha DR (1959) A bronchodilator alkaloid (vasicinone) from *Adhatoda vasica* Nees. Nature 184:1317

Atal CK (1980) Chemistry and pharmacology of vasicine—a new oxytocic and abortifacient. Raj Bandhu Industrial Co, New Delhi, p 148

Atal CK, Sharma RL, Dhar KL (1979) Chemistry and pharmacology of vasicine—a new oxytocic & abortifacient synthesis of deoxydihomo "C" vasicinone. Indian J Chem 18b:444–450

Avula B, Begum S, Ahmed S, Choudhary MI, Khan IA (2008) Quantitative determination of vasicine and vasicinone in *Adhatoda vasica* by high performance capillary electrophoresis. Pharmazie 63(1):20–22

Bagchi GD, Dwivedi PD, Haider F, Singh S, Srivastava S, Chattopadhyay (2003) Seasonal variation in vasicine content in Adhatoda species grown under north Indian plain conditions. J Med Aroma Plant Sci 25:37–40

Bhambhani S, Karwasara VS, Dixit VK, Banerjee S (2012) Enhanced production of vasicine in *Adhatoda vasica* (L.)Nees.cell culture by elicitation. Acta Physiol Plant 34(4):1571–1578

Bhartiya HP, Gupta PC (1982) A chalcone glycoside from the flowers of *Adhatoda vasica* (2′,4-dihydroxychalcone 4-glucoside). Phytochemistry 21:247

Bhat VS, Nasavatl DD, Mardikar BR (1978) *Adhatoda vasica*-an ayurvedic plant. Indian Drugs 15:62–66

Bhavyasree PG, Xavier TS (2020) Green synthesis of copper oxide/carbon nanocomposites using the leaf extract of *Adhatoda vasica* Nees, their characterization and antimicrobial activity. Heliyon 6(2):e03323

Brain KR, Thapa BB (1983) HPLC determination of Vasicine and Vasicinone in *A. vasica* Nees. J Chromatogr 258:183–188

Chakraborty A, Brantner AH (2001) Study of alkaloids from *Adhatoda vasica* Nees on their Antiinflammatory activity. Phytother Res 15:532–534

Chattopadhyay SK, Bagchi GD, Dwivedi PD, Srivastava S (2003) Process for the production of vasicine United States Patent 6676976

Chattopadhyay N, Nosál'ová G, Saha S, Bandyopadhyay SS, Flešková D, Ray B (2011) Structural features and antitussive activity of water extracted polysaccharide from *Adhatoda vasica*. Carbohydr Polym 83(4):1970–1974

Chowdhuri BK, Hirani SK (1987) HPLC study of the photochemical oxidation of Vasicine and its analogues. J Chromatogr 390:439–443

Chowdhury BK, Bhattacharyya P (1985) A further quinazoline alkaloid from *Adhatoda vasica*. Phytochemistry 24:3080–3082

Das C, Poi R, Chowdhury A (2005) HPTLC determination of vasicine and vasicinone in *Adhatoda vasica*. Phytochem Anal 16(2):90–92

Dhalwal K, Shinde VM, Mahadik KR (2010) Optimization and validation of reverse phase HPLC and HPTLC method for simultaneous quantification of Vasicine and Vasicinone in *Sida* species. J Med Plant Res 4:1289–1296

Dhankhar S, Kaur R, Ruhil S, Balhara M, Dhankhar S, Chhillar AK (2011) A review on Justicia adhatoda: a potential source of natural medicine. African J Plant Sci 5(11):620–627

Dhuley JN (1999) Antitussive effect of *Adhatoda vasica* extract on mechanical or chemical stimulation-induced coughing in animals. J Ethnopharmacol 67(3):361–365

Duraipandiyan V, Al-Dhabi NA, Balachandran C, Ignacimuthu S, Sankar C, Balakrishna K (2015) Antimicrobial, antioxidant, and cytotoxic properties of vasicine acetate synthesized from vasicine isolated from *Adhatoda vasica* L. Bio Med Res Int:1–7

Dymock W, Warden C, Hooper D (1893) Pharmacographia indica. A history of the principal drugs of vegetable origin met with in British India, vol 3. Kegan, Paul, Trench, Trubner and Co, London, pp 49–51

Gao H, Huang YN, Gao B, Li P, Inagaki C, Kawabata J (2008) Inhibitory effect on α-glucosidase by *Adhatoda vasica* Nees. Food Chem 108(3):965–972

Gibbs BF (2009) Differential modulation of IgE-dependent activation of human basophils by ambroxol and related secretolytic analogues. Int J Immunopathol Pharmacol 22(4):919–927

Gopalan S, Kulanthai K, Sadhasivam G, Pachiappan P, Rajamani S, Paramasivam D (2016) Extraction, isolation, characterization, semi-synthesis and antiplasmodial activity of Justicia adathoda leaves. Bangladesh J Pharmacol 11(4):878–885

Haq ME, Ikram M, Warsi SA (1967) Chemical composition of *Adhatoda vasica* (L.) II. Pak J Sci Ind Res 10:224–225

Hariharan V, Rangaswami S (1970) Structure of saponins A and B from the seeds of Achyranthes aspera. Phytochemistry 9(2):409–414

Ignacimuthu S, Shanmugam N (2010) Antimycobacterial activity of two natural alkaloids, vasicine acetate and 2-acetyl benzylamine, isolated from Indian shrub *Adhatoda vasica* ness. Leaves J Biosci 35(4):565–570

Ikram M, Ehsanul Haq M (1966) Estimation of vasicine from the flowering tops of *Adhatoda vasica*. Pakistan J Sci Res 18:109–110

Iyengar MA, Jambaiah KM, Kamath MS, Rao GO (1994) Studies on antiasthma Kada: a proprietary herbal combination. Indian Drugs 31:183–186

Jain MP, Sharma VK (1982) Phytochemical investigation of roots of *Adhatoda vasica*. Planta Med 46(4):250

Jayapaul K, Kishor PK, Reddy KJ (2005) Production of pyrroloquinazoline alkaloid from leaf and petiole-derived callus cultures of Adhatoda zeylanica. Vitro Cell Dev-Pl 41(5):682–685

Jha DK, Panda L, Lavanya P, Ramaiah S, Anbarasu A (2012) Detection and confirmation of alkaloids in leaves of Justicia adhatoda and bioinformatics approach to elicit its anti-tuberculosis activity. Appl Biochem Biotechnol 168(5):980–990

Jha DK, Panda L, Ramaiah S, Anbarasu A (2014) Evaluation and comparison of radical scavenging properties of solvent extracts from Justicia adhatoda leaf using DPPH assay. Appl Biochem 174(7):2413–2425

Johne S, Groeger D, Hesse M (1971) New alkaloids from *Adhatoda vasica*. Helv Chim Acta 54(3):826–834

Joshi BS, Bai Y, Puar MS, Dubose KK, Pelletier SW (1994) 1H-and 13C-NMR assignments for some pyrrolo {2, 1b} quinazoline alkaloids of *Adhatoda vasica*. J Nat Prod 57(7):953–962

Kanthale PR, Panchal VH (2014) Pharmacognostic study of *Adhatoda vasica* Nees. Biosci Discov 6(1):49–53

Karthikeyan A, Shanthi V, Nagasathaya A (2009) Preliminary phytochemical and antibacterial screening of crude extract of the leaf of *Adhatoda vasica* L. Int J Green Pharm 3(1):78–80

Khursheed A, Devender P, Ansari S (2010) Phytochemical and pharmacological investigations on Adhatoda zeylanica (Medic.): a review. Pharm J 2(12):513–519

Kumar M, Dandapat S, Kumar A, Sinha MP (2014) Pharmacological screening of leaf extract of *Adhatoda vasica* for therapeutic efficacy. Glob J Pharmacol 8(4):494–500

Lahiri PK, Prahdan SN (1964) Pharmacological investigation of Vasicinol- an alkaloid from *Adhatoda vasica* Nees. Indian J Exp Biol 2:219–223

Latha D, Prabu P, Arulvasu C, Manikandan R, Sampurnam S, Narayanan V (2018) Enhanced cytotoxic effect on human lung carcinoma cell line (A549) by gold nanoparticles synthesized from Justicia adhatoda leaf extract. Asian Pac J Trop Biomed 8(11):540

Leonard NJ, Martell MJ (1960) Laboratory realization of the SchopfOechler scheme of vasicine synthesis. Tetrahedron Lett 1:44

Liu W, Cheng XM, Wang YL, Li SP, Zheng TH, Gao YY, Wang GF, Qi SL, Wang JX, Ni JY, Wang ZT, Wang CH (2015) In vivo evaluation of the antitussive, expectorant and bronchodilating effects of extract and fractions from aerial parts of *Peganum harmala* Linn. J Ethnopharmacol 162:79–86

Liu C, Guo D, Liu L (2018) Quality transitivity and trace ability system of herbal medicine products based on quality markers. Phytomed 44:247–257

Manoharan S, Prabhakar MM (2014) *Adhatoda vasica* leaves protect cell surface glycoconjugates abnormalities during DMBA induced hamster buccal pouch carcinogenesis. Int J Pharmacogn Phytochem 6(4):817–821

Maurya S, Singh D (2010) Quantitative analysis of total phenolic content in *Adhatoda vasica* Nees extract. Int J Pharm Techol 2(4):2403–2406

Mehta DR, Naravane JS, Desai RM (1963) Vasicinone-A bronchodilator principle from *Adhatoda vasica* Nees (NO Acanthaceae). J Org Chem 28(2):445–448

Pa R, Mathew L (2012) Antimicrobial activity of leaf extracts of Justicia adhatoda L. in comparison with vasicine. Asian Pac J Trop Biomed 2(3):S1556–S1560

Pandita K, Bhatia MS, Thappa RK, Agarwal SG, Dhar KL, Atal CK (1983) Seasonal variation of alkaloids of *Adhatoda vasica* and detection of glycosides and N-oxides of vasicine and vasicinone. Planta Med 48(6):81–82

Pandiyan N, Murugesan B, Sonamuthu J, Samayanan S, Mahalingam S (2018) Facile biological synthetic strategy to morphologically aligned CeO$_2$/ZrO$_2$ core nanoparticles using Justicia adhatoda extract and ionic liquid: enhancement of its bio-medical properties. J Photochem Photobiol B 178:481–488

Panigrahi J, Gantait S, Patel IC (2017) Concurrent production and relative quantification of vasicinone from in vivo and in vitro plant parts of Malabar nut (*Adhatoda vasica* Nees). 3. Biotech 7(5):280

Parveen A, Parveen B, Parveen R, Ahmad S (2015) Challenges and guidelines for clinical trial of herbal drugs. J Pharm Bioallied Sci 7(4):329–333

Pathak RP (1970) Therapeutic guide to ayurvedic medicine.A handbook on ayurvedic medicine, vol 1. Shri Ramdayal Joshi Memorial Ayurvedic Research Institute, p 121

Patil MY, Vadivelan R, Dhanabal SP, Satishkumar MN, Elango K, Antony S (2014) Anti-oxidant, anti-inflammatory and anti-cholinergic action of *Adhatoda vasica* Nees contributes to amelioration of diabetic encephalopathy in rats: behavioral and biochemical evidences. Int J Diabetes Dev Ctries 34(1):24–31

Rajani M, Kanaki NS (2008) Chapter 19: Phytochemical standardization of herbal drugs and polyherbal formulations. In: Ramawat KG, Mérillon JM (eds) Bioactive molecules and medicinal plants. Springer, pp 349–369

Rayees S, Satti NK, Mehra R, Nargotra A, Rasool S, Sharma A, Sahu PK, Gupta VK, Nepali K, Singh G (2014) Anti-asthmatic activity of azepino [2, 1-b] quinazolones, synthetic analogues of vasicine, an alkaloid from *Adhatoda vasica*. Med Chem Res 23(9):4269–4279

Roja G, Vikrant BH, Sandur SK, Sharma A, Pushpa KK (2011) Accumulation of vasicine and vasicinone in tissue cultures of *Adhatoda vasica* and evaluation of the free radical-scavenging activities of the various crude extracts. Food Chem 126(3):1033–1038

Saha M, Bandyopadhyay PK (2017) Phytochemical screening for identification of bioactive compound and antiprotozoan activity of fresh garlic bulb over trichodinid ciliates affecting ornamental goldfish. Aquaculture 473:181–190

Saha M, Bandyopadhyay PK (2020) In vivo and in vitro antimicrobial activity of phytol, a diterpene molecule, isolated and characterized from *Adhatoda vasica* Nees.(Acanthaceae), to control severe bacterial disease of ornamental fish, Carassius auratus, caused by Bacillus licheniformis PKBMS16. Microb Pathog 15:103977

Sajeeb B, Kumar U, Md H, Bachar S (2016) Standardization of *Adhatoda vasica* nees market preparations by RP-HPLC method. Dhaka Univ J Pharm Sci 15(1):57–62

Saxena OP, Meena K, Ananta Nandtirth MA, Saxena S (1997) Identification of bio-active compounds from medicinally important plants grown *in vitro*. In: Trivedi PC (ed) Plant biotechnology recent advances. Panima Publishing Corporation, New Delhi, pp 380–388

Schopf C, Oechler F (1936) Zur Frage der Biogenese des Vasicins (Peganins). Die Synthese des Desoxyvasicins unter physiologischen Bedingungen. Justus Liebigs Ann Chem 523:1–29

Shahwar D, Raza MA, Tariq S, Riasat M, Ajaib M (2012) Enzyme inhibition, antioxidant and antibacterial potential of vasicine isolated from *Adhatoda vasica* Nees. Pak J Pharm Sci 25(3):651–656

Sharma S, Kumar M, Bhatt V, Nayal OS, Thakur MS, Kumar N, Singh B, Sharma U (2016) Vasicine from *Adhatoda vasica* as an organocatalyst for metal-free Henry reaction and reductive heterocyclization of o-nitroacylbenzenes. Tetrahedron Lett 57:5003–5008

Singh RS, Misra TN, Pandey HS, Singh BP (1991) Aliphatic gydroxyketones from *Adhatoda vasica*. Phytochemistry 30(11):3799–3801

Singh TP, Singh OM, Singh HB (2011) *Adhatoda vasica* Nees: phytochemical and pharmacological profile. Nat Prod J 1(1):29–39

Subramanya MD, Pai SR, Ankad GM, Hegde HV, Roy S, Hoti SL (2016) Simultaneous determination of vasicine and vasicinone by high-performance liquid chromatography in roots of eight Sida species. AYU 37(2):135–139

Verma N, Shukla S (2015) Impact of various factors responsible for fluctuation in plant secondary metabolites. J Appl Res Med Aromat Plants 2:105–113

Vijayanandraj S, Brinda R, Kannan K, Adhithya R, Vinothini S, Senthil K, Chinta RR, Paranidharan V, Velazhahan R (2014) Detoxification of aflatoxin B1 by an aqueous extract from leaves of *Adhatoda vasica* Nees. Microbiol Res 169(4):294–300

Wasserman HH, Kuo GH (1991) The chemistry of vicinal tricarbonyls. An efficient synthesis of (±)-vasicine. Tetrahedron Lett 32:7131–7132

Wilson DK, Shyamala G, Paulpandi M, Narayanasamy A, Siram K, Karuppaiah A, Sankar V (2020) Development and characterization of phytoniosome nano vesicle loaded with aqueous leaf extracts of *Justicia adhatoda* and *Psidium guajoava* against dengue virus (DEN-2). J Clust Sci

Yadav AK, Tangpu V (2008) Anticestodal activity of *Adhatoda vasica* extract against Hymenolepis diminuta infections in rats. J Ethnopharmacol 119(2):322–324

Yadav AK, Tandon V, Rao HSP (1992) In vitro anthelmintic efficacy of fresh tuber extract of Flemingia vestita against Ascaris suum. Fitoterapia 63:395–398

Zaynab M, Fatima M, Abbas S, Sharif Y, Umair M, Zafar MH, Bahadar K (2018) Role of secondary metabolites in plant defense against pathogens. Microb Pathog 124:198–202

Chapter 7
Pharmacological and Bioactive Basis
of *Adhatoda vasica* L. Nees

The pharmacological potential of the plant relies upon the presence of phytochemical components and their quantities. Variety of the components has their own spectrum of activities which based upon the nature of the compound and the affinity of towards the receptors. Plant *Adhatoda vasica* as being the ancient plant in use for the traditional remedies has been the choice of traditional healers. Those claims which were made in the past were on the 'Hit and Trial' basis which were supposed to be having a scientific base. This chapter includes all those approaches which were adopted to bolster the traditional claim of therapeutic practice and is based upon the scientific explorations.

7.1 Pharmacological Studies and Bioactivities

7.1.1 Metabolic Disorders

Metabolic disorders are the attribute of various causative factors where diet plan and eating habits contribute to a significant extent. The leaves extract of plant *A. vasica* has been studied to manage the metabolic disorders. Various studies suggest the enzymatic regulations could be an effective approach to control the metabolic irregularities. The present study showed the methanolic extract of leaves that was found to have highest sucrose inhibition where the sucrose. Two compounds which were elucidated (vasicine and vasicinol) through spectral analysis were found to have significant IC50 at 125 µM and 250 µM concentrations. Sucrose hydrolysing activity of α-glucosidase enzyme (mammalian intestinal enzyme) was also inhibited competitively with Ki values of 82 µM and 183 µM. Enzyme inhibition by the methanolic leaves extract of the mentioned plant plays an important role to make it more probable for diabetes to control (Gao et al. 2008). Although the present study

M. Ali, K. R. Hakeem, *Scientific Explorations of Adhatoda vasica*,
SpringerBriefs in Plant Science, https://doi.org/10.1007/978-3-030-56715-6_7

is not showing the comparable IC50 to the commercially available acarbose (α-glucosidase enzyme inhibitor) where IC50 (0.8 μM is quite lesser in comparison to the isolated compounds in the study (Gao and Kawabata 2004). Alloxane-induced diabetes was also managed by the leaves extract of vasaka and was also additionally found effective to reduce the co-morbid depression state. The findings were significant upon comparing with standard escitalopram (Gupta et al. 2014).

Alcoholic extract of the leaves of *A. vasica* showed the potential to reduce the glucose level after oral administration in rabbits and rats (Modak and Rao 1966; Dhar et al. 1968), while silver nano-formulations of the leaves extract (as capping drug) when prepared with green synthesis were shown a significant action to control the blood glucose and related parameters (Bandi and Vasundhara 2012).

Plant leaves possess the essential oils which showed the soothing effect on isolated tracheal chain of guinea pig. The effect was as smooth muscles relaxant for the ease in tracheal muscles comfort (D'Cruz et al. 1979). Ethanolic leaves extract with different doses 100, 200 and 400 mg/kg per day showed antihyperglycemic activity in rat model when the diabetes was induced through streptozotocin. Methanolic extract also exhibited the activity by countering the effect of sucrose (Hong et al. 2008). Dried extract of the vasaka leaves when combined with the beverage was also studied for the activity part and showed the anti-diabetic potential (Gedam et al. 2017).

7.1.2   Respiratory Disorders and Tuberculosis

Active constituents of *A. vasica* (vasicine and vasicinone) were undergone for in vivo and in vitro studies and found as showing bronchodilator and bronchoconstriction activities. Although vasicinone is the metabolite of vasicine but when these two are combined, they show bronchodilatory effects in vitro and in vivo conditions (Atal 1980).

Various studies pertaining to the respiratory disorders have been performed so far, but the reliable components from the natural sources attract always the researchers by keeping in mind about the safety parameters. Natural compounds always serve a significant contribution for the respiratory illness to treat where cough is one of them. Various antitussives are already employed to meet the purpose, but the current study points towards the isolation of novel compound and its effectiveness to work as antitussive remedy. It was analyzed as pectic arabinogalactan which was isolated from *A. vasica* by aqueous extraction method and was later on precipitated with ethanol. The isolated compound is a polysaccharide consisting of branched and mainly of 1,3–/1,3,6-linked galactopyranosyl and 1,5–/1,3,5-linked arabino furanosyl residues. Administration of this polysaccharide arabinogalactan with calculated dose as 50 mgkg−1 body weight through oral route revealed the inhibited frequency of number of coughs which were induced by citric acid in animal model (guinea pigs) and slightly decreased the resistance values of specific airways. That time it was first to represent in vivo antitussive activity of the aqueous extracted sugar polysaccharide and its structurally featured components. Biological investigations supported pectic arabinogalactan of *A. vasica* to display promising role in the antitussive assays. The study also helped to vindicate the claim of traditional utilities to treat the respiratory ailments in ancient times as folklore remedy (Chattopadhyay et al. 2011).

Entire plant showed the anti-allergic and anti-asthmatic potentials when administered its methanolic extract through inhalational or intragastric routes in the guinea pig at the dose 6 mg/animal or 2.5 g/kg body weight, respectively (Muller et al. 1993).

Chemical, mechanical and electrical methods were used to induce the coughing state to evaluate the antitussive activity of *A. vasica* extract which was also compared with standard codeine available as the antitussive agent. When the electrical stimulation of tracheal mucosa was done and effectiveness was compared between plant extract and codeine, it was 1:4 ratio of determination. Intravenous administration revealed the less effective outcome of the extract. In other approach where centrally induced coughing through the stimulation of vagus nerve, the effectiveness of plant extract was determined as 1:20 ratio in anaesthetized guinea pigs. The peripheral induced coughing through the mechanical stimulation of tracheal mucosal linings was also treated with the same extract potencies and found to reveal the ratio of 1:10 in terms of effectiveness. Better antitussive effect was assured when the extract was orally administered and utilized to combat the peripherally induced stimulation of cough. The alkaloids vasicinone and vasicinol as being the major phytoconstituents are believed to work as antitussive to supress the cough by acting through the cough centre or medulla as the specific neuronal site in the brain (Bucher 1958). Plant *A. vasica* extracts also possess the benefit of devoid the narcotic abuse potential, while the standard codeine bears the same (Dhuley 1999).

In folk medicinal practice, *A. vasica* was used to treat asthma and cough. The ground of the study was to be evaluated by extracting the phytoconstituents from the leaves of *A.vasica*. Different extracts (hexane, ethyl acetate and methanol) were prepared and studied for the efficacy against *Mycobacterium tuberculosis* where hexane extract showed a significant reduction in colony-forming units (CFU) at 100 μg/ml extract dose. The hexane extract was evaluated to determine the responsible components to work against these colonies and found two natural compounds as vasicine acetate and 2-acetyl benzylamine. The pure compounds were bioassayed against *Mycobacterium tuberculosis* and were found having very strong antimycobacterial activity. Vasicine acetate and 2-acetyl benzylamine significantly inhibited *M. tuberculosis* and additionally one multidrug-resistant (MDR) strain along with one sensitive strain at different dose concentrations 200 and 50 μg/ml, respectively (Ignacimuthu and Shanmugam 2010).

Semisynthetic derivatives of vasicine (benzylamines, bromhexine and ambroxol) are employed as mucolytics which was shown to have the growth inhibition of *Mycobacterium tuberculosis*. Experimental outcome favours the traditional claims where ancient saying is very popular to assure the patient of phthisis needs not to worry if vasaka plant is around (Dymock et al. 1893). As an adjuvant to the conventional therapies, these derivatives help to potentiate the effects of anti-tubercular drug rifampicin in lungs (Grange and Snell 1996).

The presence of various alkaloids and terpenes is believed to impart the action of anti-tubercular. The incidences of the bacterial growth are initiated by the biosynthesis of fatty acids which is considered to be ceased by the inactivation of enzyme β-ketoacyl-(acyl-carrier-protein) synthase III. Ultimately the growth and propagation of the bacterial strains is checked by the steps through secondary metabolites of the vasaka plant (Ahmad et al. 2009). Other extracts of the same plant were also studied for anti-tubercular action and were found effective with the revelation of considerable zone of inhibition upon comparing to the standard drugs, i.e. isoniazid and rifampicin. The plant was also suggested to be used as anti-tubercular by assuring the results through broth microdilution protocols (Jethva et al. 2016; Chidambaram and Swaminathan 2013).

The study involved the synthesis of the synthetic analogue of vasicine, which was isolated from *A. vasica*. These synthesized analogues, i.e. 10 azepino [2,1-b] quinazolone derivatives, were subjected to be evaluated for anti-asthmatic effect in an ovalbumin-induced murine model for asthma. Out of ten analogues approximately five were shown with the results by decreasing the cytokine secretions; however the more significant decrease was noted only in one synthetic derivative (Rayees et al. 2014). The alcoholic extract of the leaves also exhibited bronchoconstriction which was mediated through the acetylcholine and histamine hence beneficial as anti-asthmatic (Dangi 2015) and being the bronchodilator also support the asthmatic subject to relieve (Gandhi et al. 2015). More potent bronchodilator vasicinone as the adjuvant oxidized derivative alkaloid contributes to the anti-asthmatic approach (Nilani et al. 2009). Polyherbal formulation having the vasaka constituent (vasicine) in the standardized composition causes the mast cell stabilization and assists the animal model to recover from asthmatic symptoms in dose-dependent manner (Gohil and Mehta 2011).

Different quinazoline alkaloids from the extract of vasaka leaves were executed for docking studies to evaluate the energy and affinity to bind with the targeted sites in enzyme complex. The enzymes which are responsible for the occurrence of tuberculosis via type II fatty acid biosynthase and β-ketoacyl-acyl-carrier protein synthase III were the targets for binding in the studies. This binding affinity revealed the breakthrough to design the anti-tubercular drugs through the bioinformatics concepts. Role of natural compounds to combat the life-threatening infectious disease has already been the attention seekers for the developments of many formulations. These docking studies also showed the way to cease the initiation of diseases in early stages (Jha et al. 2012).

Polyherbal formulation Kan Jang was studied for the comparative effects with or without the standardized extract of *A. vasica*. Since the role of vasaka is known for the treatment of respiratory ailments, hence the presence and absence comparison showed the slight differences. Other standardized extracts were also effective for pulmonary and immunomodulatory efficacies (Narimaian et al. 2005). Antitussive effect of Kan Jang was also studied by adopting a very different approach with comparative and randomized double blind study tools. The standard polyherbal formulation with the fixed composition of the plant extract including vasaka was found to be very effective for upper respiratory infections (Barth et al. 2015).

Polyherbal formulation where vasaka was one of the components was investigated for antitussive action. The cough was induced to the animal model by sulphur dioxide and found a significant relief in the treated group when it was compared with the standard cough suppressant codeine drug (Harsoliya et al. 2011).

7.1.3 Antioxidant, Antimicrobial and Anti Inflammatory

The aqueous extract of the leaves possesses the potential to work against the microbial flora which was isolated from the subject infected with gingivitis (Patel and Venkata-Krishna-Bhatt 1984). The petroleum ether and alcoholic extracts of vasaka leaves were also recorded to have the antibacterial efficacies against many strains of bacteria (Sarker et al. 2009). Some study also suggested that the bimetal approach by taking silver and gold as doping tool can serve better as antibacterial vasaka extract with zinc oxide nanoparticles (Pandiyan et al. 2019). The extract of the leaves was subjected to formulate the nanoparticles of silver and gold by using cerium oxide as the loaded compound. The nanoparticles finally were tested against Gram-positive bacterial strains and were found very effective with high zone of inhibition (Nithya and Sundrarajan 2020). Previously the silver nanoparticles of the extract have been prepared and studied and investigated in 2015 by Bhumi et al.

Alkaloids as being the most bioactive components of the extract of *A. vasica* plant were evaluated for anti-inflammatory and antimicrobial activities. The constituents were already been reported in the considerable amount as vasicine and related pyrroloquinazoline compounds in the chloroform fraction of the main extract. Carrageenan was used to induce paw oedema (inflammation) in complete Freund's adjuvant (CFA) animal model, while the antimicrobial potential was assessed by microdilution method. The different doses of the phytoconstituents were found with effective variables of anti-inflammatory effect with vasicine (29 mg/kg) at 6 h after the injection of carrageenan as 59.51%, while it was maximum with another derived alkaloid vasicinone as 63.94% with 10 mg/kg of body weight when injected after 4th day of CFA injection. So less dose with highly promising anti-inflammatory effects were obtained with vasicinone which is a metabolite of main components vasicine. Antibacterial against *E. coli* and antifungal against *C. albicans* the effects were also assessed at two different doses of vasicine as 20 µg/ml and >55µg/ml respectively and were met out with significant outcome (Singh and Sharma 2013).

The leaf extracts of the plant were evaluated for the antibacterial activity where zones of inhibition were determined through the diameters against bacteria and fungi. The concentration of vasicine alkaloid was taken as 25 µg/ml and evaluated and found effective against bacteria and other pathogens. Minimum inhibitory concentration (MIC) and minimum microbicidal concentrations (MMC) were also determined against *Escherichia coli*, *Serratia marcescens* and *Pseudomonas aeruginosa*. The highest MIC was also determined against various other pathogens and was an attempt to explore and provide a useful and very substantial support for chemotherapy to control the infectious diseases (Pa and Mathew 2012). The ethanolic extract of vasaka also been recorded with promising action against multidrug resistant pathogens as well (Batool et al. 2017).

The methanolic extract of *A. vasica* leaves was subjected to be fractionated to isolate a pure compound phytol which was characterized as ($C_{20}H_{40}O$), a diterpene alcohol having a molecular weight of spectral detection m/z 297. In vivo antipathogenic potential of the isolated and characterized phytol was demonstrated the strong anti-pathogenic activity against *Bacillus licheniformis* $PKBMS_{16}$ which is responsible for mass mortality of the ornamental fish firm (Saha and Bandyopadhyay 2020).

Antioxidants are the protective agents to the proper cell functioning. The generation of the excessive reactive free radical oxygen species usually causes irreparable damage to the cellular DNA which leads to mutagenesis and finally cell death with several disorders (Simic 1988). Therefore, *A. vasica* was studied for its ability to cease the generation of reactive species and to work as antioxidant to prevent the oxidative damage to the cellular DNA. When the leaf extract of the *A. vasica* was studied for the comparative protections, it was found that the supercoiled plasmid DNA was intact. In vitro antioxidant activity was found very significant for the aqueous extract of *A. vasica*.

Antimicrobial activity was also studied against various microbial strains, i.e. (Gram +ve), namely, *S. aureus*, *B. subtilis* and (Gram −ve) *P. vulgaris* and K. pneumonia.

The plant methanolic extracts (*A. vasica* and other Indian medicinal plants) showed the significant antimicrobial activity along with the minimum inhibition concentrations (MIC) within range of 21–90 mg per ml (Khurana et al. 2010).

Cognition impairment is associated with the 'diabetic encephalopathy' which is the attribute of inflammation, cholinergic dysfunction and oxidative stresses. *A. vasica* possessing the huge range of phytochemical potential is believed to hold the promising antioxidant, anti-inflammatory, antihyperglycemic and anticholinesterase activities. Ethanolic extract of vasaka leaves was studied with three different dose potencies (100, 200 and 400 mg/kg per day) to evaluate the potential in diabetic Wistar rats where the diabetes was induced by streptozotocin in designed experiment. Different parameters were investigated in stressed and treated groups. The parameters in diabetic rat where the increased values were recorded are AchE activity, lipid peroxidation (LPO), nitrite levels and TNF-α, a pro-inflammatory cytokine, while the treated group showed the significant propitious attenuation of the behavioural and biochemical aberrations. The favourable results in the treated group were the collective outcome of antioxidant, anti-inflammatory anticholinesterase and antihyperglycemic actions (Patil et al. 2014).

Adhatoda vasica leaf extract was investigated for anti-arthritis and anti-inflammatory activities where arthritis was induced through collagen by modulating synovial tool-like receptor and facilitating the release of pro-inflammatory mediators. In the different groups, the different dose extract pattern was followed as 50, 100 and 200 mg/kg body weight and found the doses response as decreasing the arthritic index and anti-inflammatory response by reducing footpad swelling. Pro-inflammatory cytokines and other mediators were recorded as diminished. Hence the study suggested vasaka as protective tool against deleterious rheumatoid arthritis (RA). Hence it can be used in pharmacological research for novel drug development in rheumatology as well (Adhikary et al. 2016).

Shoot culture of *A. vasica* was prepared and maintained to have callus by treating with different growth regulators. Since the presence of secondary metabolites and vitamin C was already been confirmed by the previous studied, hence the callus was also proposed to be investigated by Roja et al. in 2011. The water and methanolic extracts were prepared from both the sources as parent plant and callus and finally were compared for the quantitative estimation of the alkaloids, i.e. vasicine and vasicinone, and the capacities to scavenge the free radicals generated by DPPH (2,2-diphenyl-1-picrylhydrazyl). The free radical scavenging activity of DPPH radical was found maximum in water extracts even higher in comparison with the standard analogue (water soluble) of vitamin E – Trolox (Roja et al. 2011).

Antioxidant and anti-inflammatory potential of the vasaka leaves were investigated by isolating the prime component vasicine. Asthma is the disease which carries the curative aspects of used drugs if have antioxidant and anti-inflammatory properties. The isolated vasicine showed its promising effects in the treated group of murine model of asthma where decrease in lipid peroxidation and glutathione was simultaneously observed with the significant rising of antioxidant superoxide dismutase, catalase and glutathione peroxidase (Srinivasarao et al. 2006).

The methanolic extract of vasaka was carried out, and the alkaloid fraction was separated. This fraction was studied for anti-inflammatory potential of the constituents by adopting the modified hen's egg chorioallantoic membrane test. The alkaloid fraction showed a significant activity at the calculated dose of 50 μg/pellet (Chakrabarty and Brantner 2001).

In the present study, the leaf extract of vasaka was processed to synthesize the environment-friendly silver nanoparticles. Since the plant already possesses the components to work as antibacterial, hence the nano-formulation was proved to be more efficient. The biosynthesized silver nanoparticles were also characterized by spectroscopic methods and transmission electron microscopy (TEM) analysis. The antibacterial activity of the formulation was evaluated against *Pseudomonas aeruginosa* MTCC 741 by disc diffusion method and agar cup assay method along with the serial dilution turbidity observational measurement assay (Bose and Chatterjee 2015).

Vasicine as being the main alkaloid of the *A. vasica* leaves was isolated from the ethanolic extract of the leaves. The pharmacological approach was established through its derivative vasicine acetate which was synthesized by acetylation of pure compound. The study revealed the potential of the derivative as a significant antibacterial on the basis of zone of inhibition. Different zones of inhibition were recorded as 10 mm against *E. aerogenes*, *S. epidermidis*, *P. aeruginosa* and the same derivative showed MIC values as 125 μg/ml against bacteria *M. luteus*,

S. epidermidis, E. aerogenes and *P. aeruginosa.* The derivative was also studied for free radical scavenging activity and found as 66.15% as the significant value at 1000 µg/ml. All the bioactive potentials were holding the moderate values in comparison with the vasicine (Duraipandiyan et al. 2015).

Presence of flavonoids and phenolic contents in the extract are responsible for the antioxidant or free radical scavenging activities. Since we know that the thermo labile substances are needed to be protected from heat sensitization to maintain the potential of the constituents. Theo approaches were adopted in this study where cold maceration and Soxhlet methanolic extractions OF *A. vasica* were prepared and evaluated for the free radical scavenging activities by using DPPH assays protocol. The conclusive remark of the study favoured the similar pattern of activity in both the extracts with result reproducibility. The study also supported the perception of natural antioxidant as the effective remedies to combat various detrimental diseases which have the sole causative aspects due to free radicals. This was considered as the novel approach of synergism of phytoconstituents in 2014, which further needs to be explored to isolate and determine the sole emphasis on particular metabolites (Adams Jr and Odunze 1991; Chia et al. 1984; Jha et al. 2014). Positive correlation with p value <0.05 also validated the outcome of the study in the very acceptable mode.

The alcoholic extracts of different parts might be having a variation in antibacterial activities. The leaves and roots possess the different constituents but some common as well. In the study it was shown the leaves and roots with their alcoholic extract are having the activity against *Staphylococcus aureus* and *Escherichia coli* due to maximum constituents, while water extract of both the parts showed only the activity against *S. aureus* (George et al. 1947). By the time the study patter showed the different results, Karthikeyan et al. (2009) reported in their study that the ethanolic extract of the leaves covers a wide spectrum for antimicrobial against *Staphylococcus epidermidis, Proteus vulgaris, Bacillus subtilis* and *Candida albicans.* Methanolic extract of vasaka was not found so effective against the fungal growth of *Microsporum gypseum, Trichophyton terrestre and Chrysosporium tropicum* (Quershi et al. 1997).

Ethanolic leaf extract was investigated for phytochemistry, antioxidant and free radical scavenging potential of *Adhatoda vasica.* The phytochemical analysis has already been revealed the presence of secondary metabolites (alkaloids, saponins, phenols, flavonoids, terpenoids and steroids). Antioxidant capacity was investigated by inhibiting 1, 1-diphenyl-2-picrylhydrazyl (DPPH) radical and found as 69.23%. All other free radicals were also inhibited to very significant proportions. Potent inhibitory effect was also studied on ferric ion-induced lipid peroxidation and recorded as 68.26% in bovine brain extract. These overall findings assure the biological efficiencies *A. vasica* as a potential natural source of potent antioxidants (Bajpai et al. 2015).

7.1.4 Analgesic Activity

Ethanolic and methanolic extracts of the vasaka leaves were evaluated and observed for the analgesic activities by using hot plate analgesiometer and warm water tail immersion protocols. Significant results were obtained when the writing response induced by the acetic acid was treated with the alcoholic extracts (Mulla et al. 2010; Belemkar et al. 2013). Anti-inflammatory and analgesic activities were also observed through an Ayurvedic polyherbal formulation where vasaka is one of the important components. Although the mentioned herbal formulation was designed for rheumatoid arthritis, collective impacts of anti-inflammation and analgesic were quite significant. Pure isolation of vasicine and evaluation of analgesic activity was observed and reported by Keesara and Jat in 2017(Thabrew et al. 2003).

7.1.5 Abortifacient Activity

The alkaloidal component vasicine from the plant was shown to have the comparable contractions of uterine muscles to the effect of oxytocin and other methergine. Uterotonic potential effect was not only observed in other species but human beings as well. Vasicine has the rhythmic influence on human myometrial strips irrespective of the pregnancies (Atal 1980). Uterine tonic activity was shown by the extract due to presence of the chief constituent in the extract. Uterine and abortifacient might be taken as the positive sign of acceptance when needed to deliver the baby at full term. By stimulating the uterine contraction, it provides the ease in partum (Claeson et al. 2000). Uterine stimulant or abortifacient action is also linked to the synthesis and release of prostaglandins (PGs) caused by vasicine in dose-dependent manner. The study showed its significance for the same activity; however oxytocic activity is proven as dose-dependent impact (Soni et al. 2008; Rao et al. 1982).

One study was conducted on different animal models (rats, rabbits, guinea pigs and hamsters) and revealed the potential of vasicine as uterotonic and abortifacient probably due to the synthesis and release of PG (prostaglandins). The probable significance of prostaglandin role was later on confirmed by the use of estradiol which potentiated the abortifacient effect and diminished by the administration of aspirin (a NSAID) which is known to reduce the synthesis of prostaglandins (Chandhoke 1982).

7.1.6 Hepatoprotective

Traditional claim of A. vasica to be used in liver protection was verified and assured by the study where leaf extract was administered at the dose of 50–100 mg/kg, p.o. prior to liver damage induced by d-galactosamine in rat model. It showed the significant hepatoprotection through leaf aqueous extract. And the results were compared to silymarin (dose 25 mg/kg per day) standard drug and found quite satisfactory (Bhattacharyya et al. 2005). The medicinal plants generally show the protective nature in fractionated extracts. The vasaka plant with its ethyl acetate extract showed the significant protection against the liver toxicity induced by the carbon tetrachloride (CCl_4). Injured liver showed the declined biochemical markers which get up rise to the normal values after treatment with the extract. The study was done with the considerable result and parameters. The same study was taken as the base for the other similar studies to reproduce the resulting effectiveness. Vasicinone despite bronchodilator was also reported to have hepatoprotective activity in combination with 25 mg/mg per day. Ethanolic extract of vasaka leaves having the dose of 100 to 200 mg/kg p.o. showed the significant maintenance of hepatic enzyme system which was disrupted due to toxic dose of carbon tetrachloride (Pandit et al. 2004; Ahmad et al. 2013; Kumar et al. 2015; Lisina et al. 2011; Afzal et al. 2013). Hepatoprotection was also studied by the scholars by using the available polyherbal formulation where vasaka was the important component for other applications as well (Saroj et al. 2012).

7.1.7 Antimalarial

Docking studies suggested that the adduct formulations are comparatively more effective than individual agent to fight with the diseases. The docking score of antimalarial drug chloroquine from cinchona bark is known to be more effective for the treatment of malaria when used in adduct form against mutant triosephosphate isomerase (TIM). The docking score was recorded as −11.144 kcal/mol with chloroquine-adhatodine adduct in comparison with alone chloroquine as −8.59 kcal/mol against TIM (Swain et al. 2015).

7.1.8 Wound Healing Activity

Previous study was done by Bhargava et al. (1988), and they reported about the potential of vasaka extract (alcoholic) to possess the significant effect to heal the wound in the vertebral column of a calves group. Though the chloroform extract was also studied to observe the result for comparative outcomes. The same study was repeated on rat models and found the significant to have considerable efficacy in methanolic extract of the same sample (Vinothapooshan and Sunder 2010). In a separate study where the animal model as mice was chosen to evaluate the vasaka extract ointment (1%) for wound healing properties resulted in a considerable amount of potent efficacy (Subhashini and Arunachalam 2011).

7.1.9   Anticancer Activity

Isolated compound of *Adhatoda vasica*, 2-acetyl-benzylamine was screened for different types of leukaemia cells to evaluate its anticancer activity. The leukaemia cells (NB-4, MOLM-14, CEM, IM-9, K562, Jurkat and HL-60) were subjected to for the cytotoxic studies by vasaka alkaloid 2-acetyl-benzylamine. Significant cytotoxic potential was obtained against MOLM-14 and NB-4 cells where IC50 values were 0.40 and 0.39, respectively, at 24 h upon comparison with other tested cells. Apoptosis was also confirmed by the induction of cell arrest at G0/G1 phase in NB-4 cells and G2/M phase in MOLM-14 leukaemia cells. Molecular docking and in vivo studies in xenograft mice model proved the alkaloid 2-acetyl-benzylamine from the plant could be developed and considered as a dynamic phytochemical agent to work against cancer in a very significant mode particularly for MOLM-14 category of the cancer (Balachandran et al. 2017).

Ethanolic extract of *A. vasica* leaves was investigated to evaluate the effect on cell membranes where glycoproteins play an important role in several biological functions through interactions between the cells. Altered glycoproteins may lead to some deleterious malignant transformation. The study revealed the protection of the cells by leaves extract where the surface abnormalities were triggered up by 7, 12-dimethyl benz(a)anthacene (DMBA) induction to the hamster buccal pouch to induce carcinogenesis. Study included the DMBA treatment alone for 14 weeks (thrice a week) in the buccal pouches for one group of animals. Differentiated squamous cell carcinoma was induced, and the groups were compared for the effectiveness of the extract by administering the extract dose 100 mg/kg body weight. Finally the experimental results concluded the *Adhatoda vasica* a promising natural source of protection for the cell surface abnormalities induced by the toxicant DMBA during oral carcinogenesis (Manoharan and Prabhakar 2014). Renal oxidative stress and carcinogenesis which was induced by ferric nitrilotriacetate was combated by the vasaka extract exerting its effect as anti-hyper proliferative (Jahangir and Sultana 2007). Ethyl acetate and water extract of the vasaka leaves possess the different metabolites but showed the significant activity against cell lines from cervical cancer in human being (HeLa cell line).Though it was carried out as in vitro study (Sudevan et al. 2019). Human lung epithelial cell line (HCC-827) was also made available to get conduct the anticancer efficacy of vasaka extract adopting the MTT assay. Alkaloids and their derivatives in the extracts are believed to have the role as anticancer (Chauhan et al. 2019).

Cytotoxic studies of the vasicine acetate were studied against A549 lung adenocarcinoma cancer cell line which revealed its potential as IC50 value of 2000 $\mu g/ml$. Other studies were found significant after synthetic derivatization (Duraipandiyan et al. 2015).

Vasicinone was investigated for it anti-proliferative effect against A549 lung carcinoma cells. The extract of *A. vasica* has already been shown the similar efficacies whose mechanism is supposed to be elucidated. This study showed how a quinazoline alkaloid vasicinone being the bronchodilator acts against the cancer cell lines.

Various doses of vasicinone were evaluated on A549 cell line for the effect and found the significant reduction in cell viability, lactate dehydrogenase (LDH) leakage, DNA fragmentation and altered mitochondrial potential with reduced wound healing capacities. Anti-proliferative nature of vasicinone was also revealed up by the downregulation of Bcl-2 and Fas death receptor along with upregulation of BAD, PARP and cytochrome c.

The conclusive outcome of the study is like hope to develop a new therapeutic agent as remedy to fight with life-threatening cancer against oxidative stress-induced repercussions (Dey et al. 2018).

7.1.10 Antiulcer Activity

Antiulcer activity was carried out by inducing the ulcer by pylorus ligation, ethanol and anti-inflammatory drug aspirin which was countered efficiently by the ethanolic extract of the *A. vasica* in the dose-dependent manner. The significant results were also obtained to support the antiulcer potential of the plant by using the prepared syrup having the plant as important component. The conclusive outcome was reported to mention the antiulcerogenic potential with the relief for dyspepsia for animal models (Shrivastava et al. 2006).

7.1.11 Myocardium Modulation and Cardioactive

Previous study supports the plant extract to be cardioprotective. The phytoconstituents vasicine and vasicinone in combination exhibited the significant declining in cardiac depressant condition. The racemic mixture of vasicinone (±) didn't express its efficacy, but levo form was found weakly effective to stimulate the myocardium (Atal 1980). Pure isolated vasicinone from the root was also investigated by Lahiri and Pradhan (1964) and showed the transient hypotension in cats and contraction of intestinal muscles along with the cardiac depression in guinea pigs claiming the anticholinesterase activity. This study (anti-cholinesterase) was also supported and verified by the work done by Ali et al. (2016). The presence of alkaloids and terpenoids along with the other therapeutic metabolites the the extract of Vasaka leaves was investigated and recorded to be effective as cardio protective. Since the thrombolytic features of the components are added advantages to the cardio protection (Mariyam et al. 2016). Streptokinase as being the standard thrombolytic drug was used in one study to evaluate the comparative potencies of the different extracts of the *A. vasica*, where the methanolic extract was found significant to show the desirable outcomes (Shahriar 2013; Singh et al. 2017). Methanolic extracts have their confinements to possess the different metabolites which are also dependent to the part of plant selected for the extraction. Root and stems have varied compositions of the constituents, and it was reported that the root extractives are far better to work

more thrombolytic than stem. Their efficacies in comparison with standard drug streptokinase were quite considerable (Laboni et al. 2016). Cardio protection by the extract of vasaka leaves was also reported by Roy et al (2013) and Singh et al (2011) and mentioned the presence of all types of phytochemical secondary metabolites as collective force to show the considerable result. Hossain and Hoq (2016) say about the cardiac muscle contraction due to the levo form of oxidative derivative of vasicine.

7.1.12 Anti-Alzheimer Through Acetyl Cholinesterase Inhibition

Since the Alzheimer is not completely curable, it makes not to claim 100% with any phytochemical to combat it, but the extract from *A. vasica* which contains major alkaloids as vasicine, vasicinone, anisotine and vasicole showed the significant activity as a reversible inhibitor of the specific enzyme acetyl cholinesterase (AChE) with an remarkable concentration IC50 = 294 μg/mL. Phytochemical studies revealed total alcoholic extract of *A. vasica* to yield many components where vasicine reversibly and very competitively inhibited acetyl cholinesterase enzyme with Ki value as 11.24 μM as shown in the Fig. 7.1. Contrary to the vasicine, rest of isolated components as alkaloids showed very weak inhibition on acetyl cholinesterase enzyme. The binding sequence and pattern of vasicine showed a very close resemblance to both the standard compounds galantamine and tacrine at the catalytic site. The novelty of the vasicine and its bioactive spectrum suggested the probabilities to comprise a wide pharmacological area to work on neuronal innervations which can be used anyway for the development of an efficient anti-Alzheimer drug (Ali et al. 2016).

7.1.13 Anticestodal Activity and Anthelmintic

Hymenolepis diminuta infections in different groups of rats were treated to evaluate the folk claim of *A. vasica* concerning the parasites. Anticestodal efficacy of the leaves extract was carried out by monitoring the eggs/gram (EPG) of faeces counts and worm recovery rate in terms of percentage. Finally the result showed the profound efficacy with double dose of leaf extract (800 mg/kg) against mature worms along with the reduction of EPG count as 79.57% and 16.60% worm recovery rate. These effects were significant upon comparing with 5 mg/kg single dose of standard praziquantel. The efficacy for immature worm was recorded as the reduction percentage 100% in control and 20% with 800 mg/kg dose of the extract.

The plant secondary metabolites are believed to show the collective efficacy which constitutes particularly the alkaloids, saponins, glycosides and tannins as

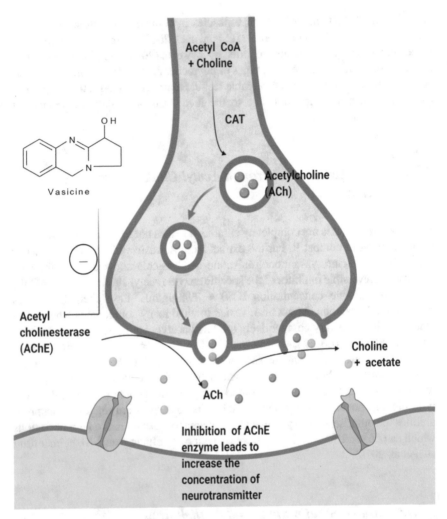

Fig. 7.1 Cholinergic hypothesis based illustration to show the role of vasicine for Alzheimer's disease. Figure is prepared through biorender (https://biorender.com/)

active to confer the antiparasitic effects (Athanasiadou and Kyriazakis 2004). The results of this study disclosed the facts about the leaves of *A. vasica* to possess anticestodal efficacious components which also validated its recommended use in folk traditions (Yadav and Tangpu 2006).

Ethanolic extract was found effective against the gastrointestinal nematodes in sheep. The evaluation was carried out by egg hatching and the assays of larval development where the dose was calculated and administered as 800 mg/kg to prove the anthelmintic capacities (Al-Shaibani et al. 2008).

The water and methanolic extract were attempted to evaluate the anthelmintic activity against the standard drug levamisole. The study showed the quite significant results for in vitro studies. Dried root powder of the plant was also found very potent against nematodes (Lateef et al. 2003).

7.1.14 Modulation Against Gamma Irradiation

Effect of gamma radiation on haematological alterations was studied to evaluate the potential of extract from vasaka leaves. Though it has been reported already the supportive views of phytoconstituents to combat the oxidative stresses in the animal models Swiss albino mice. The influences of radiation were compared at the end of study and found that the treated group of animals was quite normal even after exposure of the radiations and showed 81.25% survival within 30 days of exposure, while the radiated group without pre-treatment showed the mortality and significant decline of haematological constituents (RBCs, WBCs, Hb and Hct) values within 25 days of study along with the reduction in glutathione (GSH) content and enhancement in lipid peroxidation (LPO). However the treated group showed the significant and reliable results in the favour of better cure. The comparative symptoms were anorexia, ruffled hairs, lethargicity and diarrhoea (Kumar et al. 2015).

7.1.15 Antimutagenic Activity

Different extracts of vasaka with their corresponding polarities were studied for the antimutagenic potential and were found to have a significant potential due to the presence of highly antioxidant constituents flavonoids and phenolic components. Some other studies also supported the efficacy of the extractive components to be antimutagenic where vasicine was believed to show the effect against the mutation caused by 2-aminofluorine. The investigations were consecutive by the same group of researchers (Kaur et al. 2015; Kaur et al. 2016).

7.1.16 Thrombopoetic Potential

Reactive oxygen species (ROS) are believed to promote the maturation of megakaryocytes which may lead to produce the platelets (Motohashi et al. 2010). ROS as a by-product of superoxide anion leaked might also come from mitochondrial respiratory chain (Chen et al. 2013). The leaves extract of A. vasica was investigated to enhance the generation of mitochondrial ROS to increase the permeability of the mitochondrial membrane hence megakaryocytic maturation. The supportive potential was revealed for thrombopoietic efficiency which further needs to be evaluated to trace the proper mechanism of sequential events at molecular level (Gutti et al. 2018).

7.1.17 Larvicidal Activity

Different fractions of methanolic extract of *A. vasica* were investigated to evaluate the larvicidal activity against the bancroftian filariasis vector *Culex quinquefasciatus* and dengue vector *Aedes aegypti*. The results of the study revealed the mortality rates as quite significant at very low concentrations of methanol extract fractions. All the tested fractions were proved to have strong larvicidal activity with the calculated doses range from 100 to 250 ppm against *C. quinquefasciatus* and *A. aegypti*. The traditional claim also gets stronger by this revelation, and plant may also serve as a natural and inexpensive larvicidal agent. Methanolic extracts of the vasaka leaves were found effective against *Spodoptera littoralis*. In combination to artificial diet, the dried extract showed a significant complete mortality effect on larva. Its antifeedant nature is to serve as a biopesticide tool. Concentration variables and their spray on other growing herbs are shown to make them safe from the attacks of insects and cease the development of larva (Thanigaivel et al. 2012; Sadek 2003; Anuradha et al. 2010).

The plant is recorded to be used traditionally for insecticidal purpose in different countries where the pest control in labs and warehouses are priorities now. Other species of Culex have also been faced a death exposure by the extract in Egyptian studies. A significant potency has also been reported by Hiremath et al. (1997) against *Nilaparvata lugens* and also suggested vasaka a natural tool to replace the synthetic insecticidal products (Gangwar and Ghosh 2014; Abdelgaleil et al. 2011). Essential oils from the leaves and floral parts of the plant are determined to show the insecticidal activity against the grains stored for long time (Kokate et al. 1980). Vasicine alkaloid from the plant is considered to control a significant number of *Pseudomonas aeruginosa* in a very less concentration. The concentration revealed its potency of effectiveness (Ali et al. 2018). In the same year, the vasicine was reported as insect repellent by Yadav and Yadav (2018). Acetylated derivative of vasicine were also reported to have the larvicidal potential against *Plutella xylostella* at larva and pupa stages as well (Gabriel et al. 2014).

7.1.18 Immunomodulatory Activity

The alkaloids in the different extract of vasaka leaves are responsible for the immunomodulation in rat model. The dose was used as 400 mg/kg of the body weight, and significance of the activity was well correlated with the presence of secondary metabolites. The immunity modulation of the animal model was assured by the biochemical parameters (antibody titre values, hypersensitivity reaction and cytokine levels, etc.) as per the protocols (Singh et al. 2017; Vinothapooshan and Sundar 2011; Adhikarya et al. 2014).

7.2 Pharmacokinetics

The maximum concentration or the appearance of the vasicine in the blood is the attribute of the route adopted for the drug administration. Intramuscular route shows the significant absorption for vasicine when administered as per 20 mg/kg of the body weight. The concentration 50 mg/ml in the blood is attained irrespective of the pregnancies of the rat models and the 10 mg/ml in amniotic fluid (Atal 1980).

Maximum concentration of vasicine was obtained in 5 min in uterus when the drug was administered through intravenous route and the time extension to 10 min provided the result of peak concentration. Different routes have the different half-life span where 5–7 min is for intravenous and 1.5 h is for intramuscular, while the longest one is 2 h for subcutaneous administrations. The result reproducibility was also been vindicated by Zutshi et al. (1980).

The oral route is known to provide the very low concentrations to have in the uterus. Liver is the metabolising site for vasicine where it gets biotransformed into vasicinone and other metabolites which share the first pass effect as the an important aspect of elimination of vasicine. The reports revealed the maximum excretion of vasicine and its metabolites in urine. Different routes of administration follow the different excretion timings as intravenous for 18 h and intramuscular for 22 h, while the oral administration reveals about 18% of all the excreted product as vasicine during the first 24 h (Atal 1980).

7.3 Toxicities of *A. vasica*

Toxicity studies of vasicine were carried out on different animal models (i.e. rats and dogs) with different calculated doses for several groups with different routes of administration, but no remarkable effects were observed in any species (Atal 1980). Although the study was extended to 6 months by taking new animal models (i.e. rats and rhesus monkey), the haematological and biochemical findings were within the acceptable range of normal physiology. Even the abnormalities were not revealed in autopsy and histopathological examination (Pahwa et al. 1987). Uterotonic action of the vasicine was the base for other toxicities to be checked out, but in female rats no effect was observed even after intraperitoneal administration of vasicine at 8 or 16th day of gestation period. In guinea pigs, the vasicine when 30 mg/kg body weight was given intraperitoneally, it showed the abortion in 50% of the animals in a particular group with late pregnancy. The supportive estradiol as the pre-treatment was also given and found the enhancement of vasicine action. Some reports with the vasicine doses of 5 mg/kg to 10 mg/kg body weight show no anti-implantation effect but abortifacient and pregnancy failure after 1 week. Abortifacient effect was also observed in hamster animal model on 7th to 9th day and 10th to 12th day, while on the other hand in rabbits, abortifacient effect was reported through parenteral routes of vasicine with random variables of doses, but oral administration was not to show any remarkable change (Atal 1980). Earlier study was also supported the safety index of *A. vasica* extract by administering the dose of the extracts 325 mg/kg per day orally between 1 and 9 day of pregnancy concluding with no abortive effect in rat model (Burgos et al. 1997). Teratogenic effects were also not recorded in rats and rabbits by administering the vasicine intraperitoneally at different intervals. Some reports were supportive to abnormal happenings at 2.5 mg/kg of body weights.

The toxicity reviews of acute and general phases after repeated administration of vasicine and popular formulation Kan Jang were revealed and found no data to support any abortive effect. The formulation Kan Jang does not show any teratogenic effect or pregnancy failures (Claeson et al. 2000).

Related to the toxicity parameters and potential of the *A. vasica*, several studies have already been performed. The current study was to evaluate its nephrotoxic behaviour in vitro by using the HK-2 (Human Kidney) cell from proximal tubules. General toxicity and the cell-specific toxicity were performed and found only the short-term detrimental toxic impressions very specific to proximal tubules when the extract was exposed directly to the cells. Acute and potent toxicity of *A. vasica* under the experimental conditions envisaged a call for further study designing to make a better toxicant threshold to establish safe dosage levels. Previously the nephrotoxic effect was not determined through *A. vasica*, so novelty was studied by triggering up the toxicity by using gentamycin in vivo (Kumar et al. 2013). Although vasicine as being the major alkaloid of *A. vasica*, it is yet to be confirmed the degree of potential toxicities as a future perspective (Mossoba et al. 2016).

7.3.1 Aflatoxins Detoxification

Aqueous or hydroalcoholic extract of the plant *A. vasica* was proven to have the phytochemicals to show the potential against aflatoxin B1 (AFB1) to detoxify its impact which was later tested by TLC and ELISA. Though it was a comparative study where three different plant extracts were compared for their effectiveness. Maximum degradation of AFB1 ($\geq$98%) was revealed after incubation at 37 °C for 24 h. The detoxifying potential of vasaka extract was found declined on raising the temperature to 100 °C for 10 min or upon following the autoclaving conditions at 121 °C for 20 min. The study showed the potential to degrade 69% of the toxins within 6 h and $\geq$ 95% after 24 h of incubation which was further confirmed by liquid chromatography–mass spectrometry (LC–MS) analysis. Phytochemical analysis further confirmed the presence of alkaloids which are believed to share the detoxifying potentials. Their partially pure isolation and by preparative TLC exhibited very dynamic AFB1 detoxification. The study pointed towards the role of alkaloids to detoxify the aflatoxin; hence it opens further the ways to look for any other components to contribute for this very sensitive activity for human welfare (Vijayanandraj et al. 2014).

References

Abdelgaleil SA, Suganuma T, Kitahara K, Fujii M (2011) Insecticidal properties of extracts and phytochemicals isolated from three Egyptian plants against *Culex pipiens*. Planta Med 77(12):20

Adams JD Jr, Odunze IN (1991) Oxygen free radicals and Parkinson's disease. Free Radic Biol Med 10(2):161–169

Adhikary R, Majhi A, Mahanti S, Bishayi B (2016) Protective effects of methanolic extract of *Adhatoda vasica* Nees leaf in collagen-induced arthritis by modulation of synovial toll-like receptor-2 expression and release of pro-inflammatory mediators. J Nutr Intermed Metab 3:1–1

Adhikarya R, Majhia A, Mahantia S, Bishayia B (2014) Immunomodulatory and anti-oxidant properties of methanolic extract of *Adhatoda vasica* Nees leaf after particulate antigen stimulation in mice. J Pharm Res 8(10):1520–1537

Afzal U, Gulfraz M, Hussain S, Malik F, Maqsood S, Shah I, Mahmood S (2013) Hepatoprotective effects of *Justicia adhatoda* L. against carbon tetrachloride (CCl4) induced liver injury in Swiss albino mice. Afr J Pharm Pharmacol 7(1):8–14

Ahmad S, Garg M, Ali M, Singh M, Athar MT, Ansari SH (2009) A phyto-pharmacological overview on *Adhatoda zeylanica* Medic. syn. *A. vasica* (Linn.) Nees. Nat Prod Rad 8(5):549–554

Ahmad R, Raja V, Sharma M (2013) Hepatoprotective activity of ethyl acetate extract of *Adhatoda vasicain* Swiss albino rats. Int J Curr Res Rev 5:16–21

Ali SK, Hamed AR, Soltan MM, El-Halawany AM, Hegazy UM, Hussein AA (2016) Kinetics and molecular docking of vasicine from *Adhatoda vasica*: an acetylcholinesterase inhibitor for Alzheimer's disease. S Afr J Bot 104:118–124

Ali RH, Majeed MR, Jasim HA (2018) Determination of vasicine alkaloid efficacy as inhibitor to the activity of protease produced by a clinical isolate of *Pseudomonas aeruginosa*. Iraqi J Sci:1237–1246

Al-Shaibani RM, Phulan MS, Arijo A, Qureshi TA (2008) Ovicidal and Larvicidal properties of *Adhatoda vasica* (L.) extracts against gastrointestinal nematodes of sheep in vitro. Pak Vet J 28(2):79–83

Anuradha A, Rajan K, McConnell MS (2010) Feeding deterrence activity of *Adhatoda vasica* L. against Spodoptera litura (Fab.). J Biopest 3(1):286

Atal CK (1980) Chemistry and pharmacology of Vasicine-a new oxytocic and abortifacient. Regional Research Laboratory, Jammu- Tawi

Athanasiadou S, Kyriazakis I (2004) Plant secondary metabolites: antiparasitic effects and their role in ruminant production systems. Proc Nutr Soc 63:631–639

Bajpai VK, Agrawal P, Bang BH, Park YH (2015) Phytochemical analysis, antioxidant and anti-lipid peroxidation effects of a medicinal plant, *Adhatoda vasica*. Front Life Sci 8(3):305–312

Balachandran C, Arun Y, Sangeetha B, Duraipandiyan V, Awale S, Emi N, Ignacimuthu S, Perumal PT (2017) In vitro and in vivo anticancer activity of 2-acetyl-benzylamine isolated from *Adhatoda vasica* L. leaves. Biomed Pharmacother 93:796–806

Bandi S, Vasundhara K (2012) Green synthesis of silver nanoparticles using *Adhatoda vasica* methanolic extract and its biological activities. J Atm Mol 2(4):282–291

Barth A, Hovhannisyan A, Jamalyan K, Narimanyan M (2015) Antitussive effect of a fixed combination of Justicia adhatoda, Echinacea purpurea and Eleutherococcus senticosus extracts in patients with acute upper respiratory tract infection: a comparative, randomized, double-blind, placebo-controlled study. Phytomedicine 22(13):1195–1200

Batool R, Salahuddin H, Mahmood T, Ismail M (2017) Study of anticancer and antibacterial activities of *Foeniculum vulgare, Justicia adhatoda* and *Urtica dioica* as natural curatives. Cell Mol Biol 63(9):109–114

Belemkar S, Thakre SA, Pata MK (2013) Evaluation of anti-inflammatory and analgesic activities of methanolic extract of *Adhatoda vasica* Nees and *Mentha piperita* Linn. Inventi Rapid Ethnopharmacol 2:1–6

Bhargava MK, Singh H, Kumar A (1988) Evaluation of *Adhatoda vasica* as a wound-healing agent in buffaloes-clinical, mechanical and biochemical-studies. Indian Vet J 65(1):33–38

Bhattacharyya D, Pandit S, Jana U, Sen S, Sur TK (2005) Hepatoprotective activity of *Adhatoda vasica* aqueous leaf extract on D-galactosamine-induced liver damage in rats. Fitoterapia 76(2):223–225

Bhumi G, LingaRao M, Savithramma N (2015) Green synthesis of silver nanoparticles from the leaf extract of *Adhatoda vasica* nees. and assessment of its antibacterial activity. Asian J Pharm Clin Res 8(3):62–67

Bose D, Chatterjee S (2015) Antibacterial activity of green synthesized silver nanoparticles using Vasaka (*Justicia adhatoda* L.) leaf extract. Indian J Microbiol 55(2):163–167

Bucher K (1958) Pathophysiology and pharmacology of cough. Pharmacol Rev 10:43–58

Burgos R, Forcelledo M, Wagner H, Müller A, Hancke J, Wikman G, Croxatto H (1997) Non-abortive effect of *Adhatoda vasica* spissum leaf extract by oral administration in rats. Phytomedicine 4(2):145–149

Chakrabarty A, Brantner AH (2001) Study of alkaloids from *Adhatoda vasica* Nees on their anti-inflammatory activity. Phytother Res 15:532–534

Chandhoke N (1982) Vasicine, the alkaloid of *Adhatoda vasica*. Indian Drugs 24(9):425–426

Chattopadhyay N, Nosál'ová G, Saha S, Bandyopadhyay SS, Flešková D, Ray B (2011) Structural features and antitussive activity of water extracted polysaccharide from *Adhatoda vasica*. Carbohydr Polym 83(4):1970–1974

Chauhan N, Singh C, Gupta S (2019) Screening antibacterial efficacy and anticancer studies of *A. vasica* leaf crude extracts for formulation of potential herbal drug. J Pharmacogn Phytochem 8(1):428–434

Chen S, Meng X-F, Zhang C (2013) Role of NADPH oxidase-mediated reactive oxygen species in podocyte injury. BioMed Res Int:839761

Chia LS, Thompson JE, Moscarello MA (1984) X-ray diffraction evidence for myelin disorder in brain from humans with Alzheimer's disease. Biochim et Biophys Acta (BBA)-Biomembr 775(3):308–312

Chidambaram S, Swaminathan R (2013) Determination of anti-tubercular activity of four Indian medicinal plants against Mycobacterium tuberculosis by broth micro dilution method. Int J Pharm Sci Res 4(10):3932

Claeson UP, Malmfors T, Wikman G, Bruhn JG (2000) *Adhatoda vasica*: a critical review of ethnopharmacological and toxicological data. J Ethnopharmacol 72:1–20

D'Cruz JL, Nimbkar AY, Kokate CK (1979) Evaluation of essential oil from leaves of *Adhatoda vasica* as an airway smooth muscle relaxant. Indian J Pharm Sci 41:247

Dangi A (2015) Phytochemical screening and assessment of *Adhatoda vasica* (Leaf) for antiasthmatic activity. Panac J Pharm Sci 4(3):680–704

Dey T, Dutta P, Manna P, Kalita J, Boruah HPD, Buragohain AK, Unni B (2018) Anti-proliferative activities of vasicinone on lung carcinoma cells mediated via activation of both mitochondria-dependent and independent pathways. Biomol Ther 26(4):409

Dhar ML, Dhar MM, Dhawan BN, Mehrotra BN, Ray C (1968) Screening of indian plants for biological activity. Indian J Exp Biol 6(4):232–247

Dhuley JN (1999) Antitussive effect of *Adhatoda vasica* extract on mechanical or chemical stimulation-induced coughing in animals. J Ethnopharmacol 67(3):361–365

Duraipandiyan V, Al-Dhabi NA, Balachandran C, Ignacimuthu S, Sankar C, Balakrishna K (2015) Antimicrobial, antioxidant, and cytotoxic properties of vasicine acetate synthesized from vasicine isolated from *Adhatoda vasica* L. Bio Med Res Int:1–7

Dymock W, Warden C, Hooper D (1893) Pharmacographia Indica. A history of the principal drugs of vegetable origin met with in British India, vol 3. Kegan, Paul, Trench, Trubner and Co, London, pp 49–51

Gabriel PM, Shanmugam N, Ignacimuthu S (2014) Antifeedant activity and toxicity of two alkaloids from *Adhatoda vasica* Nees leaves against diamond back moth *Plutella xylostella* (Linn.) (Lepidoptera: Plutellidae) larvae. Arch Phytopathol Plant Protect 47(15):1832–1840

Gandhi SJ, Ahire RD, Sanap MD, Chavhan AB, Chudhari JS, Ahirrao RA (2015) Standardization of amla and vasaka. Pharma Sci Monit 6(2):50–60

Gangwar AK, Ghosh AK (2014) Medicinal uses and pharmacological activity of *Adhatoda vasica*. Int J Herb Med 2(1):88–91

Gao H, Kawabata J (2004) Importance of the B ring and its substitution on the a-glucosidase inhibitory activity of baicalein, 5,6,7- trihydroxyflavone. Biosci Biotech Bioch 68:1858–1864

Gao H, Huang YN, Gao B, Li P, Inagaki C, Kawabata J (2008) Inhibitory effect on α-glucosidase by *Adhatoda vasica* Nees. Food Chem 108(3):965–972

Gedam AM, Kshirsagar RB, Sawate AR, Patil BM (2017) Studies on physico chemical characteristics and comparative study on extraction yield of Adulasa (*Adhatoda vasica*) leaf. J Pharmacogn Phytochem 6(5):244–247

George M, Venkataraman PR, Pandalai KM (1947) Investigations on plant antibiotics. Part II A search for antibiotic substances in some Indian medicinal plants. J Sci Ind Res 6:42–46

Gohil PV, Mehta AA (2011) Evaluation of mast cell stabilizing and anti-anaphylactic activity of polyherbal formulation. Adv Biol Res 5(6):304–308

Grange JM, Snell NJC (1996) Activity of bromhexine and ambroxol, semi-synthetic derivatives of vasicine from the Indian shrub *Adhatoda vasica*, against *Mycobacterium tuberculosis in vitro*. J Ethnopharmacol 50(1):49

Gupta D, Radhakrishnan M, Kurhe Y (2014) Effects of *Adhatoda vasica* leaf extract in depression co-morbid with alloxan-induced diabetes in mice. Int J Green Pharm 8(2):97–104

Gutti U, Komati JK, Kotipalli A, Saladi RG, Gutti RK (2018) Justicia adhatoda induces megakaryocyte differentiation through mitochondrial ROS generation. Phytomedicine 43:135–139

Harsoliya MS, Patel VM, Singh S, Pathan JK (2011) Anti tussive effect of multi uses medicinal plants on sulfur dioxide gas induced cough reflex in mice. J Pharm Res 4(11):4123–4125

Hiremath G, Ahn YJ, Kim SI (1997) Insecticidal activity of Indian plant extracts against *Nilaparvata lugens* (*Homoptera Delphacidae*). Appl Entomol Zool 32(1):159–166

Hong G, Na HY, Bo G, Peng L, Chika I, Jun K (2008) Inhibitory effect on a-glucosidase by *Adhatoda vasica* Nees. Food Chem 108:965–972

Hossain MT, Hoq MO (2016) Therapeutic use of *Adhatoda vasica*. Asian J Med Biol Res 2(2):156–163

Ignacimuthu S, Shanmugam N (2010) Antimycobacterial activity of two natural alkaloids, vasicine acetate and 2-acetyl benzylamine, isolated from Indian shrub *Adhatoda vasica* ness. leaves. J Biosci 35:565–570

Jahangir T, Sultana S (2007) Tumor promotion and oxidative stress in ferric nitrilo triacetate mediated renal carcinogenesis: protection by *Adhatoda vasica*. Toxicol Mech Method 17(7):421–430

Jethva K, Bhatt D, Zaveri M (2016) In-vitro anti-tuberculosis activity of selected ethnomedicinal plants. Int J Herb Med 4(4):126–128

Jha DK, Panda L, Lavanya P, Ramaiah S, Anbarasu A (2012) Detection and confirmation of alkaloids in leaves of Justicia adhatoda and bioinformatics approach to elicit its anti-tuberculosis activity. Appl Biochem 168(5):980–990

Jha DK, Panda L, Ramaiah S, Anbarasu A (2014) Evaluation and comparison of radical scavenging properties of solvent extracts from Justicia adhatoda leaf using DPPH assay. Appl Biochem 174(7):2413–2425

Karthikeyan A, Shanthi V, Nagasathaya A (2009) Preliminary phytochemical and antibacterial screening of crude extract of the leaf of *Adhatoda vasica* L. Int J Green Pharm 3(1):78–80

Kaur A, Kaur D, Arora S (2015) Evaluation of antioxidant and antimutagenic potential of *Justicia adhatoda* leaves extract. Afr J Biotechnol 14(21):1807–1819

Kaur A, Katoch D, Singh B, Arora S (2016) Seclusion of vasicine-an quinazoline alkaloid from bioactive fraction of *Justicia adhatoda* and its antioxidant, antimutagenic and anticancerous activities. J Global Biosci 5(4):3836–3850

Keesara BR, Jat RK (2017) Isolation and characterization of Vasicine from *Adhatoda vasica* (Adusa). Int J Res 6(2):2590–2596

Khurana R, Karan R, Kumar A, Khare SK (2010) Antioxidant and antimicrobial activity in some Indian herbal plants: protective effect against free radical mediated DNA damage. J Plant Biochem Biotechnol 2:229–233

Kokate CK, Tipnis HP, Gonsalves LX (1980) Anti-insect and juvenile hormone mimicking activities of essential oils of *Adhatoda vasica*, *Piper longum* and *Cyperus rotundus*. In 4. Asian symposium on medicinal plants and spices, Bangkok (Thailand)

Kumar A, Kumari SN, D'Souza P, Bhargavan D (2013) Evaluation of renal protective activity of Adhatoda zeylanica (medic) leaves extract in wistar rats. Nitte Univ J Health Sci 3(4):45

Kumar M, Dandapat S, Sinha MP (2015) Hepatoprotective activity of *Adhatoda vasica* and *Vitex negundo* leaf extracts against carbon tetrachloride induced hepatotoxicity in rats. Adv Biol Res 9(4):242–246

Laboni FR, Batul UK, Uddin JA, Karim NU, Labu ZK, Rashid MH (2016) Antioxidant, cytotoxic, sedative and anti-diuretic effect of stem and root extracts of Basak, *Adhatoda vasica*. World J Sci Engineering 1(1):31–38

Lahiri PK, Pradhan SN (1964) Pharmacological investigation of vasicinol -an alkaloid from *Adhatoda vasica* Nees. Indian J Exp Biol 2:219–222

Lateef M, Iqbal Z, Khan MN, Akhtar MS, Jabbar A (2003) Anthelmintic activity of Adhatoda vesica roots. Int J Agric Biol 5(1):86–90

Lisina KV, Ragavendran P, Sophia D, Rajamanikandan S, Sindhu T, Durgapriya D, Gopalakrishnan VK (2011) A comparative study of *Justicia adhatoda*, *Mimosa pudica* and *Vitex negundo* agaist hepatoprotective activity in albino rats *in vivo* evaluation. Pharmacology Online 1:481–491

Manoharan S, Prabhakar MM (2014) *Adhatoda vasica* leaves protect cell surface glycoconjugates abnormalities during DMBA induced hamster buccal pouch carcinogenesis. Int J Pharmacogn Phytochem 6(4):817–821

Mariyam AA, Isaac RS, Praseetha PK (2016) Phytochemical profile and pharmacognistic properties of adhatoda species: a review. Adv Sci Eng Med 8(9):669–675

Modak AT, Rao MR (1966) Hypoglycemic activity of a non nitrogenous principle from the leaves of *Adhatoda vasica* nees. Indian J Pharm 28:105–106

Mossoba ME, Flynn TJ, Vohra SN, Wiesenfeld PL, Sprando RL (2016) In vitro exposure of Adhatoda zeylanica to human renal cells lacks acute toxicity. Toxico Rep 3:15–20

Motohashi H, Kimura M, Fujita R, Inoue A, Pan X, Takayama M, Katsuoka F, Aburatani H, Bresnick EH, Yamamoto M (2010) NF-E2 domination over Nrf2 promotes ROS accumulation and megakaryocytic maturation. Blood, J Am Soc Hematol 115(3):677–686

Mulla WA, More SD, Jamge SB, Pawar AM, Kazi MS, Varde MR (2010) Evaluation of antiinflammatory and analgesic activities of ethanolic extract of roots *Adhatoda vasica* Linn. Int J Pharmtech Res 2(2):1364–1368

Muller A, Antus S, Bittinger M, Kaas A, Kreher B, Neszmelyi A, Stuppner H, Wagner H (1993) Chemistry and pharmacology of antiasthmatic *Galphimia glauca*, *Adhatoda* 6*asica*, and *Picrorhiza kurrooa*. Planta Med 59(A):586–587

Narimaian M, Badalyan M, Panosyan V, Gabrielyan E, Panossian A, Wikman G (2005) Randomized trial of a fixed combination (KanJang) of herbal extracts containing *Adhatoda vasica*, *Echinacea purpurea* and *Eleutherococcus senticosus* in patients with upper respiratory tract infections. Phytomedicine 12(8):539–547

Nilani P, Kasthuribai N, Duraisamy B, Dhamodaran P, Ravichandran S, Ilango K, Suresh B (2009) *In vitro* antioxidant activity of selected antiasthmatic herbal constituents. Anc Sci Life 28(4):3

Nithya P, Sundrarajan M (2020) Ionic liquid functionalized biogenic synthesis of AgAu bimetal doped CeO2 nanoparticles from *Justicia adhatoda* for pharmaceutical applications: antibacterial and anti-cancer activities. J Photoch Photobio B 202:111706

Pa R, Mathew L (2012) Antimicrobial activity of leaf extracts of Justicia adhatoda L. in comparison with vasicine. Asian Pac J Trop Biomed 2(3):S1556–S1560

Pahwa GS, Zutschi U, Atal CK (1987) Chronic toxicity studies with vasicine from *Adhatoda vasica* Nees in rats and monkeys. Indian J Exp Biol 25:467–470

Pandit S, Sur TK, Jana U, Debnath PK, Sen S, Bhattacharyya D (2004) Prevention of carbon tetrachloride-induced hepatotoxicity in rats by *Adhatoda vasica* leaves. Indian J Pharmacol 36(5):312

Pandiyan N, Murugesan B, Arumugam M, Sonamuthu J, Samayanan S, Mahalingam S (2019) Ionic liquid-A greener templating agent with *Justicia adhatoda* plant extract assisted green synthesis of morphologically improved Ag-Au/ZnO nanostructure and it's antibacterial and anticancer activities. J Photoch Photobio B 198:111559

Patel VK, Venkata-Krishna-Bhatt H (1984) In vitro study of antimicrobial activity of *Adhatoda* 6*asica* Linn. (leaf extract) on gingival inflammation-a preliminary report. Indian J Med Sci 38:70–72

Patil MY, Vadivelan R, Dhanabal SP, Satishkumar MN, Elango K, Antony S (2014) Anti-oxidant, anti-inflammatory and anti-cholinergic action of *Adhatoda vasica* Nees contributes to amelioration of diabetic encephalopathy in rats: behavioral and biochemical evidences. Int J Diabetes Dev Ctries 34(1):24–31

Quershi S, Rai MK, Agrawal SC (1997) *In vitro* evaluation of inhibitory nature of extracts of 18-plant species of Chhindwara against 3- keratinophilic fungi. Hindus Antibiot Bull 39:56–60

Rao MNA, Krishnan S, Jain MP, Anand KK (1982) Synthesis of vasicine and vasicinone derivatives for oxytoxic and bronchodilatory activity. Indian J Pharmaceutic Sci 44:151–152

Rayees S, Satti NK, Mehra R, Nargotra A, Rasool S, Sharma A, Singh G (2014) Anti-asthmatic activity of azepino [2, 1-b] quinazolones, synthetic analogues of vasicine, an alkaloid from *Adhatoda vasica*. Med Chem Res 23(9):4269–4279

Roja G, Vikrant BH, Sandur SK, Sharma A, Pushpa KK (2011) Accumulation of vasicine and vasicinone in tissue cultures of *Adhatoda vasica* and evaluation of the free radical-scavenging activities of the various crude extracts. Food Chem 126(3):1033–1038

Sadek MM (2003) Antifeedant and toxic activity of *Adhatoda vasica* leaf extract against Spodoptera littoralis (Lep., Noctuidae). J Appl Entomol 127(7):396–404

Saha M, Bandyopadhyay PK (2020) In vivo and in vitro antimicrobial activity of phytol, a diterpene molecule, isolated and characterized from *Adhatoda vasica* Nees.(Acanthaceae), to control severe bacterial disease of ornamental fish, Carassius auratus, caused by Bacillus licheniformis PKBMS16. Microb Pathog 15:103977

Sarker AK, Ahamed K, Chowdhury JU, Begum J (2009) Characterization of an expectorant herbal basak tea prepared with *Adhatoda vasica* leaves. Bangladesh J Sci Indus Res 44(2):211–214

Saroj BK, Mani D, Mishra SK (2012) Scientific validation of polyherbal hepatoprotective formulation against paracetamol induced toxicity. Asian Pac J Trop Biomed 2(3):S1742–S1746

Shahriar M (2013) Phytochemical screenings and thrombolytic activity of the leaf extracts of *Adhatoda vasica*. Int J Sci Tech The Experimen 7(4):438–441

Shrivastava N, Srivastava A, Banerjee A, Nivsarkar M (2006) Antiulcer activity of *Adhatoda vasica Nees*. J Herb Pharmacother 6:43–49

Simic MG (1988) Mechanisms of inhibition of free-radical processes in mutagenesis and carcinogenesis. Mutat Res Fund Mol M 202(2):377–386

Singh B, Sharma RA (2013) Anti-inflammatory and antimicrobial properties of pyrroloquinazoline alkaloids from *Adhatoda vasica* Nees. Phytomedicine 20(5):441–445

Singh SK, Patel JR, Dangi A, Bachle D, Kataria RK (2017) A complete over review on *Adhatoda vasica* a traditional medicinal plants. J Med Plant 5(1):175–180

Soni S, Anandjiwala S, Patel G, Rajani M (2008) Validation of different methods of preparation of *Adhatoda vasica* leaf juice by quantification of total alkaloids and vasicine. Indian J Pharm Sci 70(1):36

Srinivasarao D, Jayaraj IA, Jayraaj R, Prabha ML (2006) A study on antioxidant and anti-inflammatory activity of Vasicine against lung damage in rats. Indian J Allergy Asthma Immunol 20(1):1–7

Subhashini S, Arunachalam KD (2011) Investigations on the phytochemical activities and wound healing properties of *Adhatoda vasica* leave in Swiss albino mice. African J Plant Sci 5(2):133–145

Sudevan S, Parasivam R, Sundar S, Velauthan H, Ramasamy V (2019) Investigation of anti-inflammatory and anti-cancer activity of Justicia adathoda metabolites. Pak J Pharm Sci 32(4):1555–1561

Swain SS, Sahu MC, Padhy RN (2015) In silico attempt for adduct agent (s) against malaria: combination of chloroquine with alkaloids of *Adhatoda vasica*. Comput Meth Prog Bio 122(1):16–25

Thabrew MI, Dharmasiri MG, Senaratne L (2003) Anti-inflammatory and analgesic activity in the polyherbal formulation Maharasnadhi Quathar. J Ethnopharmacol 85(2–3):261–267

Thanigaivel A, Chandrasekaran R, Revathi K, Nisha S, Sathish-Narayanan S, Kirubakaran SA, Senthil-Nathan S (2012) Larvicidal efficacy of *Adhatoda vasica* (L.) Nees against the bancroftian filariasis vector Culex quinquefasciatus say and dengue vector Aedes aegypti L. in in vitro condition. Parasitol Res 110(5):1993–1999

Vijayanandraj S, Brinda R, Kannan K, Adhithya R, Vinothini S, Senthil K, Chinta RR, Paranidharan V, Velazhahan R (2014) Detoxification of aflatoxin B1 by an aqueous extract from leaves of *Adhatoda vasica* Nees. Microbiol Res 169(4):294–300

Vinothapooshan G, Sundar K (2011) Immunomodulatory activity of various extracts of *Adhatoda vasica* Linn. in experimental rats. Afr J Pharm Pharmacol 5(3):306–310

Vinothapooshan G, Sunder K (2010) Hepatoprotective activity of *Adhatoda vasica* leaves against carbo tetrachloride induced toxicity. Pharmacology Online 2:551–558

Yadav AK, Tangpu V (2006) Anticestodal activity of *Adhatoda vasica* extract against Hymenolepis diminuta infections in rats. J Ethnopharmacol 119(2):322–324

Yadav S, Yadav VK (2018) Ethnomedicinal value and pharmacognosy of the member of Acanthaceae: *Adhatoda vasica* (Linn.). Asian Pac J Health Sci 5(2):40–43

Zutshi U, Rao PG, Soni A, Gupta OP, Atal CK (1980) Absorption and distribution of vasicine, a novel uterotonic. Planta Med 40:373–377

Chapter 8
Economic Importance and Formulations

Herbal formulations of *Adhatoda vasica* are worldwide available to combat the various ailments in a very natural way. Core tablets containing 300 mg of dried powder of *A. vasica* were also attempted to formulate and further were evaluated with all standard parameters. These core tablets were prepared by direct compression method where ingredients were as vasaka powder, croscarmellose sodium (2.5–7.5%), lactose, starch, magnesium stearate and aerosol. The drug diluents and super disintegrating agents were passed through sieve no. 40 and then were blended in a plastic container (Sarfaraz and Joshi 2014). Traditional and folklore uses of the medicinal plants not only inspire to formulate the solid dosage forms but gel preparations as well. The conceptual grounds took various other plants to blend their extract in the calculated proportions with the alcoholic extract of *A. vasica* to prepare the polyherbal gel formulation which was also evaluated and studied for its stability and finally was screened for antibacterial efficacies (Bhinge et al. 2019). The extracts which contributed to make this vasaka gel preparation effective were from *Azadirachta indica, Piper betle, Pongamia pinnata* and *Ocimum tenuiflorum*.

The economic aspects of the herbal drug industries are so wide, and annual turnover is quite significant support to contribute the country's economy. Since the natural drug *A. vasica* is indigenous to the Asian regions, its availability and formulations are mainly linked to Asian herbal drug industries. Other countries (i.e. Germany, Sweden, etc.) also formulate some products of vasaka to cure the diseases in natural ways. Some selective brands of the *A. vasica* preparations are covered up in Table 8.1. Indian manufacturing companies have various brands of formulations in the market which include modern preparations, i.e. Geri forte, Dap syrup, Kofol syrup, Swasonil syrup, Vasa syrup, Koflet syrup, Zelly paste, Bresol, Styplon wet, HomRop wet, Evecare, Lukol, Diakof, etc. While modern Ayurveda preparations

Table 8.1 Branded
formulations of *A. vasica*
with the countries belong

Formulations	Country	References
Kada	India	Iyengar et al. (1994)
Femoforte	India	Shete (1993)
Salus tuss	Germany	Rote Liste (1977)
Kan jang	Sweden	Färnlöf (1998)
Spirote	Sweden	Färnlöf (1998)
Linkus syrup	Pakistan	Sheikh et al. (2014)

Table 8.2 Important polyherbal formulations having *A. vasica* as the main ingredients

Formulation	References
The contents of **Kan Jang mixture** per 100 ml – extraction of *Echinacea* corresponding to 10 g of *purpurea*, radix et herba dried material, extraction of *Adhatoda vasica*, corresponding to 5 g of folia dried material, extraction of *Acanthopanax* corresponding to 5 g of *senticosus*, radix dried material, extraction of *Glycyrrhiza* corresponding to 5 g of *glabra*, radix-dried material and vitamin C 1 g	Claeson et al. (2000)
Composition for 10 ml of Linkus syrup	Sheikh et al. (2014)
Adhatoda vasica (dry leaves Extract) – 600 mg	
Glycyrrhyza glabra (dry root extract) – 75 mg	
Piper longum (dry fruit extract) – 100 mg	
Viola odorata (dry fruit extract) – 25 mg	
Hyssopus officinalis (dry leaves extract – 50 mg	
Alpinia galangal (dry root extract) – 50 mg	
Cordia latifolia (dry fruit extract – 100 mg	
Althea officinalis (dry extract of flower) – 100 mg	
Ziziphus jujuba (dry fruit extract) – 100 mg	
Onosma bracteatum (dry leaves extract) – 100 mg	

include chywanprash Rasayana Arkey cough, Breth Easi, Atrisor, Dama Buti, Mukof, etc. Two important formulation of *A. vasica* (Kan Jang mixture and Linkus syrup) are tabulated with specified compositions for 100 ml and 10 ml doses in Table 8.2.

References

Bhinge SD, Bhutkar MA, Randive DS, Wadkar GH, Kamble SY, Kalel PD, Kadam SS (2019) Formulation and evaluation of polyherbal gel containing extracts of Azadirachta indica, *Adhatoda vasica*, Piper betle, Ocimum tenuiflorum and Pongamia pinnata. Marmara Pharm J 23(1):44–54

Claeson UP, Malmfors T, Wikman G, Bruhn JG (2000) *Adhatoda vasica*: a critical review of ethnopharmacological and toxicological data. J Ethnopharmacol 72(1–2):1–20

Färnlöf A (1998) Naturla̋kemedel och Naturmedel. Ha̋lsokostra°dets Förlag, Stockholm. 109 p 132

Iyengar MA, Jambaiah KM, Kamath MS, Rao GO (1994) Studies on an antiasthma Kada: a proprietary herbal combination, part I. Clinical study. Indian Drugs 31:183–186

Rote Liste (1977) Bundesverband der Pharmazeutischen Industrie e.V, Frankfurt a.M

Sarfaraz M, Joshi VG (2014) Development and evaluation of enteric coated herbal drug delivery system for treatment of asthma. IOSR J Pharm 4(12):34–46

Sheikh ZA, Zahoor A, Khan SS, Usmanghani K (2014) Design, development and phytochemical evaluation of a poly herbal formulation linkus syrup. Chin Med 5:104–112

Shete AB (1993) Femiforte, indigenous herbomineral formulation in the management of nonspecific leucorrhoea. Doctor's News 5:13–14

Chapter 9
Patent Status for Phyto-pharmacological Aspects

The beneficial biological activities of *Adhatoda vasica* are known to herbal healers since ancient time though its bioactive phytochemical constituent vasicine, a quinazoline alkaloid, was isolated and characterized in the early twentieth century (Sen and Ghose 1924). The other important constituents such as vasicol, vasicinone, vasicinol, deoxyvasicinone, etc. were discovered afterwards (Amin and Mehta 1959; Gupta et al. 1977). The beneficial medicinal effects of *Adhatoda vasica* extract are attributed to vasicine; hence analytical methods were developed to quantify vasicine, other related phytochemicals and total alkaloids in various *Adhatoda vasica* preparations (Chauhan and Kimothi 1999; Srivastava et al. 2001; Soni et al. 2008). Bagchi et al. in 2003 reported that the vasicine content (1.22–2.57%) in *Adhatoda* species grown in North India vary from season to season (Bagchi et al. 2003). Vasicine has been used extensively as one of the main ingredients in several herbal formulations for treating diseases related to respiratory and female reproductive systems. Therefore, attempts were made by industries and scholars to improve the production and isolation process of vasicine from the plant. Bagchi et al. (2007) and Chattopadhyay et al. (2003, 2002) obtained the patents for their inventions which were related to the improved process for the production of vasicine from the leaves of *Adhatoda vasica* formula. The procedure described the preparation and concentration of alcoholic extract of dried and powdered leaves at room temperature followed by treatment with an aqueous organic acid for 2–24 h along with stirring. The aqueous acidic layer was separated, neutralized with an alkaline solution and then extracted with an organic solvent which upon evaporation yielded an amorphous residue. The treatment of this residue with organic solvent(s) produced the pure biologically active vasicine.

Several polyherbal formulations containing *Adhatoda vasica* or vasicine for the treatment of cold, cough, laryngitis, bronchitis or asthma and the process related to their preparations have also been patented. Synthetic derivatives of vasicine or its

M. Ali, K. R. Hakeem, *Scientific Explorations of Adhatoda vasica*, SpringerBriefs in Plant Science, https://doi.org/10.1007/978-3-030-56715-6_9

analogues have also been investigated to develop drugs effective in respiratory diseases (Bruce and Kumar 1968).

Doshi et al. in 2002 patented the process and preparation of a polyherbal cough formulation. The formulation contained extracts of 17 Ayurvedic plants in different proportions. *Adhatoda vasica* (10–25%) was one of the major plant extracts in the formulation beside *Ocimum sanctum, Glycyrrhiza glabra* and *Curcuma longa* (Doshi et al. 2002). Another cough syrup formulation containing plant extracts of *Adhatoda, Hedychium* and *Curcuma* was patented by Singh and his group in 2005 (Singh et al. 2005).

Murali and co-workers obtained a patent for their invention related to the preparation process and composition of a polyherbal formulation developed for the treatment of bronchial asthma (Murali 2003). The herbal preparation contained 17 medicinal plant extracts and was claimed to have synergistic activity in bronchial asthma. It has been published under the Patent Cooperation Treaty (PCT). An invention involving formulation of a polyherbal cough lozenge was patented by Katiyar and his group under PCT in 2006 (Katiyar et al. 2006). The developed pharmaceutical formulation was effective in the treatment and management of cold symptoms such as cough and sore throat. The invention also included the processes related to the preparation, qualitative and quantitative analysis of all phytoconstituents present in the formulation. *Adhatoda vasica* Nees because of its bronchodilatory and expectorant action was used along with seven other indigenous herbs to invent an anti-asthmatic drug (Asmakure) based on multi-dimensional clinical approach (MDCA). The preparation was shown to afford immediate relief in symptoms of acute bronchitis and also caused depression of vegal terminals (Chatterjee 2003). It was claimed that asmakure not only offers complete curative treatment for asthmatic patients but also provides full immunity against common cold. The various phytochemicals present in the preparation included vasicine, monocyclic sesquiterpenes zingiberine, piperine, etc. The piperine was claimed to enhance the bio-availability of the formulation.

An invention related to chemical composition of a shampoo containing no or low concentration of pyrethrum was developed to kill the head lice. The head lice formulation in liquid or gel form containing 5–30% of *Adhatoda* and *Stemona* was shown by Munro et al., to increase the kill rate of lice (Munro et al. 2004).

Palpu et al. in 2006 successfully obtained a patent for the processing and composition of a herbal nutritious chocolate as a nutrition source. The chocolate was claimed to boost energy, vitality and exert useful anti-fatigue and anti-stress actions. The invention of a nutritious chocolate comprised several herbs with antioxidant, antianxiety and immunostimulant activities. The formulation was prepared by mixing having extracts/juices of *Adhatoda vasica* (4–7%) with *Tinospora cordifolia, Glycyrrhiza glabra, Madhuca indica* and *Cassia occidentalis* with acceptable additives such as starch, lactose, sugar, gum acacia and known lubricants (Palpu et al. 2006). Similar invention by the same research group of Council of Scientific Industrial Research, India was granted a US patent in 2007 for the development of a solid or semisolid nutritious polyherbal formulation containing decoction of *Adhatoda vasica* (Palpu et al. 2007).

The herbal formulation invention related to the low cost method and treatment of coccidiosis by using a mixture of five plant extracts was patented by Oinam in 2012 under PCT. The preparation used admixture of leaves of *Adhatoda vasica* (1–2 parts) with *Allium odorum* (2–4 parts), *Allium sativum* (1–2 parts), *Tinospora cordifolia* (2–3 parts) and *Tridax procumbens* (1–2 parts) in a suitable vehicle was claimed to be free from side effects and provided relief to the poultry from coccidiosis (Oinam 2012). An invention related to an improved process for preparing a medicinal smoking cigarette purely using Ayurvedic plants was awarded an Indian patent to Patel in 1992 (Patel 1992).

In addition to the pharmaceutical formulations of *Adhatoda vasica* Nees, two Japanese and one Chinese patents pertaining to its use in cosmetic preparations are also found in data base. The patents were granted for *Adhatoda vasica* containing cholesteric liquid crystals cosmetic preparation for having improved whitening, antioxidation and hyaluronidase inhibiting activities vis a vis an improved moisture holding property. The patents were filed by Mikimoto Pharmaceutical Co. Ltd. alone or and jointly with Nanba Tsuneo [http://www.pfc.org.in].

The details of awarded patents related to *Adhatoda vasica* or its phytoconstituents are presented in Table 9.1.

Table 9.1 A list of patents granted to *Adhatoda vasica* or its phytoconstituents

S. No	Title	Inventors	Patent no and year
1.	Process for the production of vasicine	Chattopadhyay SK et al.	US6676976, 21 Mar 2002
2.	Process for the production of vasicine	Chattopadhyay SK et al.	WO/2003/080618, 2003.
3.	Improved method for the production of vasicine	Bagchi GD et al.	DE60214438T2, 2007-01-04
4.	Herbal cough formulations and process for the preparation thereof	Doshi M et al.	US20030228383A1, 2002-10-07
5.	Herbal formulation comprising extracts of *Adhatoda*, *Hedychium* and *Curcuma* as cough syrup	Singh R et al.	WO2005077393A1, 2005-08-25
6.	Polyherbal composition for the treatment of bronchial asthma and the process	Murali PM	WO2003055558A1, WIPO (PCT), 2003-07-10
7.	Herbal formulations as cough lozenge	Katiyar CK et al.	WO2006067600A2, WIPO (PCT), 2006-06-29
8.	New anti-asthmatic drug (asmakure) from indigenous herbs to cure the disease asthma	Chatterjee TK	CA2503669A1, 2003
9.	Head lice formulation	Munro D et al.	US20040005344A1, 2004-01-08
10.	Herbal nutritious composition and its processing	Palpu P et al.	WO2006061850A1, WIPO (PCT), 2006-06-15

(continued)

Table 9.1 (continued)

S. No	Title	Inventors	Patent no and year
11.	Herbal nutritious chocolate formulation and process for preparation thereof	Palpu P et al.	US7247322B2, 2007-07-24
12.	A herbal formulation to treat coccidiosis	Oinam ID	WO2012131731A1, WIPO (PCT), 2012-10-04
13.	An improved process for preparing Ayurvedic medicinal plant-based Ayurvedic plant smoking cigarette	Patel AH	A61K035078, 26-12-92
14.	Compound Tibetan medicine preparation for treating fatty liver and alcoholic liver, expelling gall bladder and intestine roundworms and detoxifying	Chinese patent (in Chinese)	CN105125766A, 2015-12-09
15.	Tibetan medicine compound preparation for treating eye diseases	Chinese patent (in Chinese)	CN105079320A, 2015-11-25
16.	Cosmetic	Shimomura K, Ueda K	JPH07157420A, 1993
17.	Cosmetics	Nanba Tsuneo and Mikimoto Pharmaceut Co. Ltd	JPH07118135A, 1993
18.	*Adhatoda vasica* skin-bright astringent	Gong L, Chinese patent in Chinese	CN101732194A, 2010

References

Amin AH, Mehta DR (1959) A bronchodilator alkaloid (vasicinone) from *Adhatoda vasica* Nees. Nature 184:1317

Bagchi GD, Dwivedi PD, Haider F, Singh S, Srivastava S, Chattopadhyay SK (2003) Seasonal variation in vasicine content in *Adhatoda* Species grown under North Indian plain conditions. J Med Arom Plant Sci 25:37–40

Bagchi GD, Chattopadhyay SK, Dwivedi PD, Srivastava S (2007) Improved method for the production of vasicine. DE60214438T2

Bruce RA, Kumar V (1968) The effect of a derivative of vasicine on bronchial mucus. Br J Clin Pract 22:289–292

Chatterjee TK (2003) New anti-asthmatic drug (asmakure) from indigenous herbs to cure the disease asthma. CA2503669A1

Chattopadhyay SK, Bagchi GD, Dwivedi PD, Srivastava S (2002) Process for the production of vasicine. US6676976

Chattopadhyay SK, Bagchi GD, Dwivedi PD, Srivastava S (2003) An improved process for the production of vasicine. WO/2003/080618

Chauhan SK, Kimothi GP (1999) Development of HPLC method for vasicine and vasicinone in *Adhatoda vasica* Nees. Indian J Nat Prod 15:21–24

Doshi M, Vasavada S, Joshi M, Mody S (2002) Herbal cough formulations and process for the preparation thereof. US20030228383A1

Gupta OP, Sharma M, Ghatak BJR, Atal CK (1977) Pharmacological investigations of vasicine and vasicinone-the alkaloids of *Adhatoda vasica*. Indian J Med Res 66:680–691

IPR, Herbs related Patents. Available at http://www.pfc.org.in/fac/feb01.pdf. Accessed 31 Mar 2020

Katiyar CK, Padiyar A, Singh R, Kumar R, Kanaujia A , Sharma NK (2006) Herbal formulations as cough lozenge. WO2006067600A2/WIPO (PCT)

Munro D, Munro J, Bone K (2004) Head lice formulation. US20040005344A1

Murali PM (2003) Polyherbal composition for the treatment of bronchial asthma and the process. WO2003055558A1/WIPO (PCT)

Oinam ID (2012) A herbal formulation to treat coccidiosis. WO2012131731A1/WIPO (PCT)

Palpu P, Rawat AKS, Rao CV, Ojha SK, Reddy GD (2006) Herbal nutritious composition and its processing. WO2006061850A1/WIPO (PCT)

Palpu P, Rawat AKS, Rao CV, Ojha SK, Reddy GD (2007) Herbal nutritious chocolate formulation and process for preparation thereof. US7247322B2

Patel AH (1992) An improved process for preparing Ayurvedic medicinal plant based Ayurvedic plant smoking cigarette. A61K035078

Sen JN, Ghose TO (1924) Alkaloid from leaves of *Adhatoda vasica*. J Indian Chem Soc 1:315

Singh R, Padiyar A, Kanaujia A, Sharma NK (2005) Herbal formulation comprising extracts of Adhatoda, Hedychium and Curcuma as cough syrup. WO2005077393A1

Soni S, Anandjiwala S, Patel G, Rajani M (2008) Validation of different methods of preparation of *Adhatoda vasica* leaf juice by quantification of total alkaloids and vasicine. Indian J Pharm Sci 70:36–42

Srivastava S, Verma RK, Gupta MM, Singh SC, Kumar S (2001) HPLC determination of vasicine in *Adhatoda vasica* with PDA detector. J Liq Chromatogr Relat Technol 24:153–159

Chapter 10
Conclusion and Future Perspectives

Plants are known to provide the natural chemo-constituents of diverse categories where secondary metabolites are of enormous pharmaceutical importance. The present book concludes the overall potential of the Asian remedy *Adhatoda vasica* L. Nees which is known by different names in different regional languages. The conclusive part has been summarized along with the future possibilities of research avenues to come. These are as follows:

- The overview of biotechnological approaches adopted for *Adhatoda vasica* plant suggest that the alkaloidal part of the plant can be enhanced further by adopting the skilled measures under controlled conditions which would not be bound for the seasonal confinements. The source of active pharmaceutical metabolites is key factor to decide the fate of biogenesis and synthesis. The enhanced production of secondary metabolites (vasicine, vasicinone, etc.) can be managed by using the bio elicitors after thorough examination of comparative differences when bio elicitors are not used. Although the reliable tools and approaches have already been used for the purpose and have been covered in the book, the future priorities would be embarked upon the better production of callus in less time to produce the adequate amount of pharmaceutical targets. Constant challenges and refinements would certainly help the researchers to save time by exploring the hidden perspectives of scientific novelty.
- The plant *A. vasica* was also studied by Roja et al. in 2011 for quantitative estimation and comparison of the phytochemical compounds (vasicine, vasicinone, etc.) in shoot culture and parent plant. The water extract showed the maximum content and antioxidant potential upon comparing to the methanolic extract of the same and standard water-soluble vitamin E analogue. The future prospects of the study suggest the research projects must be designed to explore the remaining metabolites in the extracts and the bioactivities pertaining to them (Roja et al. 2011).

M. Ali, K. R. Hakeem, *Scientific Explorations of Adhatoda vasica*,
SpringerBriefs in Plant Science, https://doi.org/10.1007/978-3-030-56715-6_10

95

- The leaves extract of A. *vasica* showed the enzyme-guided fraction which needs to be explored further for the bioactive enzymatic inhibitory role of two main compounds (vasicine and vasicinol) to supress hyperglycaemia (postprandial) for the diabetic patient (Gao et al. 2008). Through a critical review and toxicological summaries, Claeson et al. (2000) concluded that the scientific validations don't support any remarkable toxicities of vasicine in particular or the extract from the plant A. *vasica* for human beings; hence it can be considered safe. However, the study and technical approaches are now advanced in the last 20 years, so some corrective measures can also be adopted to evaluate the toxic parameters again by considering the geographical and climatic changes into account. Though vasicine-based studies would reflect the reproducible results, the whole extract might be having some unexpected compositions.
- Chloroform extract of A. *vasica* have the most effective pyrroloquinazoline alkaloids (vasicine, etc.) isolated and were shown the significant anti-inflammatory, anti-bacterial, and anti-fungal activities. The novel bioactivities and constant efforts to explore the hidden potential of the plant would provide greater opportunities to ponder insight the medicinal values and therapeutic utilities of pyrroloquinazoline alkaloids (Singh and Sharma 2013). The broad-spectrum antimicrobial and antifungal effect of leaves extract support the probabilities to get utilize the plant extract to develop the novel chemotherapeutic agents to work against the infectious diseases (Pa and Mathew 2012).
- Antitussive effect of the A. *vasica* was found promising in different animal models. The extract does not carry the narcotic abuse potential, while the standard drugs, i.e. codeine, possess which was studied to compare in the present study. The plant extract sought the further needs to work upon to determine any dose-dependent ease of effectiveness and any other toxicological aspect if present (Dhuley 1999).
- The detoxification of the aflatoxins was carried out by A. *vasica* extract and studied well to check out the phytochemicals responsible for the activity. This was done first time in the year 2014, and further studies are still required to be undertaken for structural elucidation of the aflatoxin detoxifying principle in A. *vasica* if it exists and to determine the degraded products of aflatoxin B1 (AFB1) in the various other edible items to protect the human lives from probable carcinogenicity (Vijayanandraj et al. 2014).
- Docking studies are the novel approaches to make the individual pharmaceutical component more effective in adduct formulation. The alkaloid adhatodine was found more effective with chloroquine to work against the malarial mutant TIM. Thus, the adduct formulation could be further explored and pursued in pharmacological studies for the verification of the masked but favourable facts recorded therein (Swain et al. 2015).
- The safety margin and multi-dynamic safe vasicine alkaloid from the plant with its reversible binding pattern to the acetyl cholinesterase enzyme suggests it to be the promising leading compound to combat the Alzheimer disease. Further studies may support to make it safer and secure for other neurological innervations where the disorders are to be cured (Ali et al. 2016).

- The isolation of the polysaccharides such as pectic arabinogalactan and others involves a few easy inexpensive, safe and easy steps which is always considered as an advantage to develop the drugs for respiratory disorders. Further studies to isolate and purify other polysaccharides could be of great importance to execute the sequential development of various other promising drugs (Chattopadhyay et al. 2011).
- Anticestodal activity was carried out and found very significant at double dose of the leaf extract 800 mg/kg which suggested the presence of secondary metabolites of great importance. The metabolites further needs to be isolated for future explorations of phytochemical presence and efficacy. The traditional claims are supported by the study results and envisaged further strategies for anthelmintic potentials (Yadav and Tangpu 2006).
- Vasicine acetate and 2-acetyl benzylamine were isolated from hexane extract of vasaka leaves and showed very profound activity against *Mycobacterium tuberculosis* which could further be studied to evaluate and develop the new drugs for this hazardous and very infectious pulmonary disorder (Ignacimuthu and Shanmugam 2010).
- Protection of the cells is an approach to stop the cellular damage from the oxidative stress. *A. vasica*, methanolic extracts of the leaves were studied and showed the significant protection of supercoiled plasmid DNA which was further confirmed by the DNA isolation and proper real-time polymerase chain reaction (PCR) protocols. The result developed the interest to explore the actual components in the methanolic extract of the *A. vasica* as the part of future perspectives (Khurana et al. 2010).
- Vasaka extract for the antibacterial evaluation in nano-form is the effective approach to meet the requirement. The researchers suggested the green synthesized Ag nanoparticles to make the formulations more economical and promising option for the antibacterial purpose in the field of agricultural sciences and pharmaceutical priorities (Bose and Chatterjee 2015).
- Pure compound vasicine has its spectrum of activities, while the synthetic approaches suggest that the derivative of the vasicine as vasicine acetate after acetylation may bring a huge and very significant or moderate change in antioxidant, antibacterial and anticancer activities (Duraipandiyan et al. 2015).
- Vasaka alkaloids vasicine and vasicinone were attempted to be evaluated for cardioprotective activity by Atal (1980). The effects in combinations or on individual basis were observed which were not very significant except few. The literature survey pertaining to the cardioprotection or toxicity was not been carried out. The presence of several known and unknown compounds in different extracts might be a new overwhelming study after 30 years. It must be taken into consideration.
- Different fractions of methanolic extract of vasaka were investigated and found larvicidal in nature at very low concentration. This study suggests further evaluation of different extracts and fractions at different Rf values to achieve the significant mortality rates against other vectors of diseases (Thanigaivel et al. 2012).

- Last but not least the suggestive measures for future prospects favour to have a keen observance on current research on this Asian remedy *Adhatoda vasica*. During the studies we faced many ambiguous findings pertaining to the phyto-constituents and their relation to the different activities; however, those were dose-dependent effects which sometimes lack the reproducibility by other researchers. We hope the hidden potential of the remedy would be continuing to get explored for the human welfare.

References

Ali SK, Hamed AR, Soltan MM, El-Halawany AM, Hegazy UM, Hussein AA (2016) Kinetics and molecular docking of vasicine from *Adhatoda vasica*: an acetylcholinesterase inhibitor for Alzheimer's disease. S Afr J Bot 104:118–124

Atal CK (1980) Chemistry and pharmacology of vasicine-a new oxytocic and abortifacient, Regional Research Laboratory, Jammu- Tawi

Bose D, Chatterjee S (2015) Antibacterial activity of green synthesized silver nanoparticles using Vasaka (*Justicia adhatoda* L.) leaf extract. Indian J Microbiol 55(2):163–167

Chattopadhyay N, Nosál'ová G, Saha S, Bandyopadhyay SS, Flešková D, Ray B (2011) Structural features and antitussive activity of water extracted polysaccharide from *Adhatoda vasica*. Carbohydr Polym 83(4):1970–1974

Claeson UP, Malmfors T, Wikman G, Bruhn JG (2000) *Adhatoda vasica*: a critical review of eth-nopharmacological and toxicological data. J Ethnopharmacol 72:1–20

Dhuley JN (1999) Antitussive effect of *Adhatoda vasica* extract on mechanical or chemical stimulation-induced coughing in animals. J Ethnopharmacol 67(3):361–365

Duraipandiyan V, Al-Dhabi NA, Balachandran C, Ignacimuthu S, Sankar C, Balakrishna K (2015) Antimicrobial, antioxidant, and cytotoxic properties of vasicine acetate synthesized from vasi-cine isolated from *Adhatoda vasica* L. Bio Med Res Int:1–7

Gao H, Huang YN, Gao B, Li P, Inagaki C, Kawabata J (2008) Inhibitory effect on α-glucosidase by *Adhatoda vasica* Nees. Food Chem 108(3):965–972

Ignacimuthu S, Shanmugam N (2010) Antimycobacterial activity of two natural alkaloids, vasi-cine acetate and 2-acetyl benzylamine, isolated from Indian shrub *Adhatoda vasica* Ness. leaves. J Biosci 35:565–570

Khurana R, Karan R, Kumar A, Khare SK (2010) Antioxidant and antimicrobial activity in some Indian herbal plants: protective effect against free radical mediated DNA damage. J Plant Biochem Biot (2):229–233

Pa R, Mathew L (2012) Antimicrobial activity of leaf extracts of Justicia adhatoda L. in compari-son with vasicine. Asian Pac J Trop Biomed 2(3):S1556–S1560

Roja G, Vikrant BH, Sandur SK, Sharma A, Pushpa KK (2011) Accumulation of vasicine and vasicinone in tissue cultures of *Adhatoda vasica* and evaluation of the free radical-scavenging activities of the various crude extracts. Food Chem 126(3):1033–1038

Singh B, Sharma RA (2013) Anti-inflammatory and antimicrobial properties of pyrroloquinazo-line alkaloids from *Adhatoda vasica* Nees. Phytomedicine 20(5):441–445

Swain SS, Sahu MC, Padhy RN (2015) In silico attempt for adduct agent (s) against malaria: combination of chloroquine with alkaloids of *Adhatoda vasica*. Comput Meth Prog Biol 22(1):16–25

Thanigaivel A, Chandrasekaran R, Revathi K, Nisha S, Sathish-Narayanan S, Kirubakaran SA, Senthil-Nathan S (2012) Larvicidal efficacy of *Adhatoda vasica* (L.) Nees against the bancroft-ian filariasis vector Culex quinquefasciatus Say and dengue vector Aedes aegypti L. in in vitro condition. Parasitol Res 110(5):1993–1999

Vijayanandraj S, Brinda R, Kannan K, Adhithya R, Vinothini S, Senthil K, Chinta RR, Paranidharan V, Velazhahan R (2014) Detoxification of aflatoxin B1 by an aqueous extract from leaves of *Adhatoda vasica* Nees. Microbiol Res 169(4):294–300

Yadav AK, Tangpu V (2006) Anticestodal activity of *Adhatoda vasica* extract against Hymenolepis diminuta infections in rats. J Ethnopharmacol 119(2):322–324